T0210616

Introduction to Logic

Third Edition

Synthesis Lectures on Computer Science

The Synthesis Lectures on Computer Science publishes 75-150 page publications on general computer science topics that may appeal to researchers and practitioners in a variety of areas within computer science.

© Springer Nature Switzerland AG 2022

Reprint of original edition © Morgan & Claypool 2017

All rights reserved. No part of this publication may be reproduced, stored in a retrieval system, or transmitted in any form or by any means—electronic, mechanical, photocopy, recording, or any other except for brief quotations in printed reviews, without the prior permission of the publisher.

Introduction to Logic, Third Edition
Michael Genesereth and Eric J. Kao

ISBN: 978-3-031-00673-9   paperback
ISBN: 978-3-031-01801-5   ebook

DOI 10.1007/978-3-031-01801-5

A Publication in the Springer series
SYNTHESIS LECTURES ON COMPUTER SCIENCE

Lecture #8
Series ISSN
Synthesis Lectures on Computer Science
Print 1932-1228   Electronic 1932-1686

Introduction to Logic

Third Edition

Michael Genesereth
Stanford University

Eric J. Kao
VMware, Inc.

SYNTHESIS LECTURES ON COMPUTER SCIENCE #8

ABSTRACT

This book is a gentle but rigorous introduction to Formal Logic. It is intended primarily for use at the college level. However, it can also be used for advanced secondary school students, and it can be used at the start of graduate school for those who have not yet seen the material.

The approach to teaching logic used here emerged from more than 20 years of teaching logic to students at Stanford University and from teaching logic to tens of thousands of others via online courses on the World Wide Web. The approach differs from that taken by other books in logic in two essential ways, one having to do with content, the other with form.

Like many other books on logic, this one covers logical syntax and semantics and proof theory plus induction. However, unlike other books, this book begins with Herbrand semantics rather than the more traditional Tarskian semantics. This approach makes the material considerably easier for students to understand and leaves them with a deeper understanding of what logic is all about.

In addition to this text, there are online exercises (with automated grading), online logic tools and applications, online videos of lectures, and an online forum for discussion. They are available at

http://intrologic.stanford.edu//.

KEYWORDS

Formal Logic, Symbolic Logic, Propositional Logic, Herbrand Logic, Relational Logic, deduction, reasoning, Artificial Intelligence

Contents

Preface

This book is a first course in Formal Logic. It is intended primarily for use at the college level. However, it can also be used for advanced secondary school students, and it can be used at the start of graduate school for those who have not yet seen the material.

There are just two prerequisites. The book presumes that the student understands sets and set operations, such as union, intersection, and so forth. It also presumes that the student is comfortable with symbolic manipulation, as used, for example, in solving high-school algebra problems. Nothing else is required.

The approach to teaching Logic used here emerged from more than 10 years of experience in teaching the logical foundations of Artificial Intelligence and more than 20 years of experience in teaching Logic for Computer Scientists. The result of this experience is an approach that differs from that taken by other books in Logic in two essential ways, one having to do with content, the other with form.

The primary difference in content concerns that semantics of the logic that is taught. Like many other books on Logic, this one covers first-order syntax and first-order proof theory plus induction. However, unlike other books, this book starts with Herbrand semantics rather than the more traditional Tarskian semantics.

In Tarskian semantics, we define an interpretation as a universe of discourse together with a function (1) that maps the object constants of our language to objects in a universe of discourse and (2) that maps relation constants to relations on that universe. We define variable assignments as assignments to variables. We define the semantics of quantified expressions as variations on variable assignments, saying, for example, that a universally quantified sentence is true for a given interpretation if and only if it is true for every variation of the given variable assignment. It is a mouthful to say and even harder for students to understand.

In Herbrand semantics, we start with the object constants, function constants, and relation constants of our language; we define the Herbrand base (i.e. the set of all ground atoms that can be formed from these components); and we define a model to be an arbitrary subset of the Herbrand base. That is all. In Herbrand semantics, an arbitrary logical sentence is logically equivalent to the set of all of its instances. A universally quantified sentence is true if and only if all of its instances are true. There are no interpretations and no variable assignments and no variations of variable assignments.

Although both approaches ultimately end up with the same deductive mechanism, we get there in two different ways. Deciding to use Herbrand semantics was not an easy to choice to make. It took years to get the material right and, even then, it took years to use it in teaching Logic. Although there are some slight disadvantages to this approach, experience suggests that the advantages significantly outweigh those disadvantages. This approach is considerably easier for students to understand and leaves them with a deeper understanding of what Logic is all about. That said, there are some differences between Herbrand semantics and Tarskian semantics that some educators and theoreticians may find worrisome.

First of all, Herbrand semantics is not compact—there are infinite sets of sentences that are inconsistent while every finite subset is consistent. The upshot of this is that there are infinite sets of sentences where we cannot demonstrate unsatisfiability with a finite argument within the language itself. Fortunately, this does not cause any practical difficulties, since in all cases of practical interest we are working with finite sets of premises.

One significant deficiency of Herbrand semantics vis a vis Tarskian semantics is that with Herbrand semantics there are restrictions on the cardinality of the worlds that can be axiomatized. Since there is no external universe, the cardinality of the structures that can be axiomatized is equal to the number of ground terms in the language. (To make things easy, we can always choose a countable language. We can even choose an uncountable language, though doing so would ruin some of the nice properties of the logic. On the positive side, it is worth noting that in many practical applications we do not care about uncountable sets. Although there are uncountably many real numbers, remember that there are only countably many floating point numbers.) More significantly, recall that the Lowenheim-Skolem Theorem for Tarskian semantics assures us that even with Tarskian semantics we cannot write sentences that distinguish models of different infinite cardinalities. So, it is unclear whether this restriction has any real significance for the vast majority of students.

Herbrand semantics shares most important properties with Tarskian semantics. In the absence of function constants, the deductive calculus is complete for all finite axiomatizations. In fact, the calculus derives the exact same set of sentences. When we add functions, we lose this nice property. However, we get some interesting benefits in return. For one, it is possible with Herbrand semantics (with functions) to finitely axiomatize arithmetic. As we know from Godel, this is not possible in a first-order language with Tarskian semantics. The downside is that we lose completeness. However, it is nice to know that we can at least define things, even though we cannot prove them. Moreover, as mentioned above, we do not actually lose any consequences that we are able to deduce with Tarskian semantics.

That's all for what makes the content of this book different from other books. There is also a difference in form. In addition to the text of the book in print and online, there are also online exercises (with automated grading), some online Logic tools and applications, online videos of lectures, and an online forum for discussion.

The online offering of the course began with an experimental version early in the 2000s. While it was moderately successful, we were at that time unable to combine the online materials and tools and grading program with videos and an online forum, and so we discontinued the experiment. Recently, it was revived when Sebastian Thrun, Daphne Koller, and Andrew Ng created technologies for comprehensive offering online courses and began offering highly successful online courses of their own. With their technology and the previous materials, it was easy to create a comprehensive online course in Logic. And this led to completion of this book.

Thanks also to Pat Suppes, Jon Barwise, John Etchemendy, David-Barker Plummer, and others at the Stanford Center for the Study of Language and Information for their pioneering work on online education in Logic. *Language, Proof, and Logic* (LPL) in particular is a wonderful introduction to Logic and is widely used around the world. Although there are differences between that volume and this one in theory (especially semantics) and implementation (notably the use here of browser-based exercises and applications), this volume is in many ways similar to LPL. In particular, this volume shamelessly copies the LPL tactic of using online worlds (like Tarski's World) as a teaching tool for Logic.

And thanks as well to the thousands of students who over the years have had to endure early versions of this material, in many cases helping to get it right by suffering through experiments that were not always successful. It is a testament to the intelligence of these students that they seem to have learned the material despite multiple bumbling mistakes on our part. Their patience and constructive comments were invaluable in helping us to understand what works and what does not.

Finally, we need to acknowledge the enormous contributions of a former graduate student—Tim Hinrichs. He is a co-discoverer of many of the results about Herbrand semantics, without which this book would not have been written.

Michael Genesereth and Eric J. Kao
October 2016

CHAPTER 1

Introduction

1.1 INTRODUCTION

Logic is one of the oldest intellectual disciplines in human history. It dates back to Aristotle. It has been studied through the centuries by people like Leibniz, Boole, Russell, Turing, and many others. And it is still a subject of active investigation today.

We use Logic in just about everything we do. We use the language of Logic to state observations, to define concepts, and to formalize theories. We use logical reasoning to derive conclusions from these bits of information. We use logical proofs to convince others of our conclusions.

And we are not alone! Logic is increasingly being used by computers—to prove mathematical theorems, to validate engineering designs, to diagnose failures, to encode and analyze laws and regulations and business rules.

Logic is also becoming more common at the interface between man and machine, in "logic-enabled" computer systems, where users can view and edit logical sentences. Think, for example, about email readers that allow users to write rules to manage incoming mail messages—deleting some, moving others to various mailboxes, and so forth based on properties of those messages. In the business world, eCommerce systems allow companies to encode price rules based on the product, the customer, the date, and so forth.

Moreover, Logic is sometimes used not just by users in communicating with computer systems but by software engineers in building those systems (using a programming methodology known as *logic programming*).

This chapter is an overview of Logic as presented in this book. We start with a discussion of possible worlds and illustrate the notion in an application area known as Sorority World. We then give an informal introduction to the key elements of Logic—logical sentences, logical entailment, and logical proofs. We then talk about the value of using a formal language for expressing logical information instead of natural language. Finally, we discuss the automation of logical reasoning and some of the computer applications that this makes possible.

1.2 POSSIBLE WORLDS

Consider the interpersonal relations of a small sorority. There are just four members—Abby, Bess, Cody, and Dana. Some of the girls like each other, but some do not.

Figure 1.1 shows one set of possibilities. The checkmark in the first row here means that Abby likes Cody, while the absence of a checkmark means that Abby does not like the other girls (including herself). Bess likes Cody too. Cody likes everyone but herself. And Dana also likes the popular Cody.

	Abby	Bess	Cody	Dana
Abby			✓	
Bess			✓	
Cody	✓	✓		✓
Dana			✓	

Figure 1.1: One state of Sorority World.

Of course, this is not the only possible state of affairs. Figure 1.2 shows another possible world. In this world, every girl likes exactly two other girls, and every girl is liked by just two girls.

	Abby	Bess	Cody	Dana
Abby	✓		✓	
Bess		✓		✓
Cody	✓		✓	
Dana		✓		✓

Figure 1.2: Another state of Sorority World.

As it turns out, there are quite a few possibilities. Given four girls, there are sixteen possible instances of the *likes* relation—Abby likes Abby, Abby likes Bess, Abby likes Cody, Abby likes Dana, Bess likes Abby, and so forth. Each of these sixteen can be either true or false. There are 2^{16} (65,536) possible combinations of these true-false possibilities, and so there are 2^{16} possible worlds.

1.3   LOGICAL SENTENCES

Let's assume that we do not know the likes and dislikes of the girls ourselves but we have informants who are willing to tell us about them. Each informant knows a little about the likes and dislikes of the girls, but no one knows everything.

Here is where Logic comes in. By writing *logical sentences*, each informant can express exactly what he or she knows—no more, no less. For our part, we can use the sentences we have been told to draw conclusions that are *logically entailed* by those sentences. And we can use *logical proofs* to explain our conclusions to others.

Figure 1.1 shows some logical sentences pertaining to our sorority world. The first sentence is straightforward; it tells us directly that Dana likes Cody. The second and third sentences tell us what is not true without saying what is true. The fourth sentence says that one condition holds or another but does not say which. The fifth sentence gives a general fact about the girls Abby likes. The sixth sentence expresses a general fact about Cody's likes. The last sentence says something about everyone.

Dana likes Cody.
Abby does not like Dana.
Dana does not like Abby.
Bess likes Cody and Dana.
Abby likes everyone that Bess likes.
Cody likes everyone who likes her.
Nobody likes herself.

Figure 1.3: Logical sentences describing Sorority World.

Sentences like these constrain the possible ways the world could be. Each sentence divides the set of possible worlds into two subsets, those in which the sentence is true and those in which the sentence is false. Believing a sentence is tantamount to believing that the world is in the first set. Given two sentences, we know the world must be in the intersection of the set of worlds in which the first sentence is true and the set of worlds in which the second sentence is true. Ideally, when we have enough sentences, we know exactly how things stand.

Effective communication requires a language that allows us to express what we know, no more and no less. If we know the state of the world, then we should write enough sentences to communicate this to others. If we do not know which of various ways the world could be, we need a language that allows us to express only what we know. The beauty of Logic is that it gives us a means to express incomplete information when that is all we have and to express complete information when full information is available.

1.4   LOGICAL ENTAILMENT

Logical sentences can sometimes pinpoint a specific world from among many possible worlds. However, this is not always the case. Sometimes, a collection of sentences only partially constrains the world. For example, there are four different worlds that satisfy the sentences in Figure 1.3, viz. the ones shown in Figure 1.4.

	Abby	Bess	Cody	Dana
Abby			✓	
Bess			✓	
Cody	✓	✓		✓
Dana			✓	

	Abby	Bess	Cody	Dana
Abby		✓	✓	
Bess			✓	
Cody	✓	✓		✓
Dana			✓	

	Abby	Bess	Cody	Dana
Abby			✓	
Bess			✓	
Cody	✓	✓		✓
Dana		✓	✓	

	Abby	Bess	Cody	Dana
Abby		✓	✓	
Bess			✓	
Cody	✓	✓		✓
Dana		✓	✓	

Figure 1.4: Four states of Sorority World.

Even though a set of sentences does not determine a unique world, it is often the case that some sentences are true in every world that satisfies the given sentences. A sentence of this sort is said to be a *logical conclusion* from the given sentences. Said the other way around, a set of sentences *logically entails* a conclusion if and only if every world that satisfies the sentences also satisfies the conclusion.

What can we conclude from the bits of information in Figure 1.3? Quite a bit, as it turns out. For example, it must be the case that Bess likes Cody. Also, Bess does not like Dana. There are also some general conclusions that must be true. For example, in this world with just four girls, we can conclude that everybody likes somebody. Also, everyone is liked by somebody.

> *Bess likes Cody.*
> *Bess does not like Dana.*
> *Everybody likes somebody.*
> *Everybody is liked by somebody.*

Figure 1.5: Conclusions about Sorority World.

One way to check whether a set of sentences logically entails a conclusion is to examine the set of all worlds in which the given sentences are true. For example, in our case, we notice that, in every world that satisfies our sentences, Bess likes Cody, so the statement that Bess likes Cody is a logical conclusion from our set of sentences.

1.5 LOGICAL PROOFS

Unfortunately, determining logical entailment by checking all possible worlds is impractical in general. There are usually many, many possible worlds; and in some cases there can be infinitely many.

The alternative is *logical reasoning*, viz. the application of reasoning rules to derive logical conclusions and produce *logical proofs*, i.e., sequences of reasoning steps that leads from *premises* to *conclusions*.

The concept of proof, in order to be meaningful, requires that we be able to recognize certain reasoning steps as immediately obvious. In other words, we need to be familiar with the reasoning "atoms" out of which complex proof "molecules" are built.

One of Aristotle's great contributions to philosophy was his recognition that what makes a step of a proof immediately obvious is its form rather than its content. It does not matter whether you are talking about blocks or stocks or sorority girls. What matters is the structure of the facts with which you are working. Such patterns are called *rules of inference*.

As an example, consider the reasoning step shown below. We know that all Accords are Hondas, and we know that all Hondas are Japanese cars. Consequently, we can conclude that all Accords are Japanese cars.

All Accords are Hondas.
All Hondas are Japanese.
Therefore, all Accords are Japanese.

Now consider another example. We know that all borogoves are slithy toves, and we know that all slithy toves are mimsy. Consequently, we can conclude that all borogoves are mimsy. What's more, in order to reach this conclusion, we do not need to know anything about borogoves or slithy toves or what it means to be mimsy.

All borogoves are slithy toves.
All slithy toves are mimsy.
Therefore, all borogoves are mimsy.

What is interesting about these examples is that they share the same reasoning structure, viz. the pattern shown below.

All x are y.
All y are z.
Therefore, all x are z.

The existence of such reasoning patterns is fundamental in Logic but raises important questions. Which patterns are correct? Are there many such patterns or just a few?

Let us consider the first of these questions. Obviously, there are patterns that are just plain wrong in the sense that they can lead to incorrect conclusions. Consider, as an example, the (faulty) reasoning pattern shown below.

All x are y.
Some y are z.
Therefore, some x are z.

Now let us take a look at an instance of this pattern. If we replace x by *Toyotas* and y by *cars* and z by *made in America*, we get the following line of argument, leading to a conclusion that happens to be correct.

All Toyotas are cars.
Some cars are made in America.
Therefore, some Toyotas are made in America.

On the other hand, if we replace x by *Toyotas* and y by *cars* and z by *Porsches*, we get a line of argument leading to a conclusion that is questionable.

All Toyotas are cars.
Some cars are Porsches.
Therefore, some Toyotas are Porsches.

What distinguishes a correct pattern from one that is incorrect is that it must *always* lead to correct conclusions, i.e., they must be correct so long as the premises on which they are based are correct. As we will see, this is the defining criterion for what we call *deduction*.

Now, it is noteworthy that there are patterns of reasoning that are sometimes useful but do not satisfy this strict criterion. There is inductive reasoning, abductive reasoning, reasoning by analogy, and so forth.

Induction is reasoning from the particular to the general. The example shown below illustrates this. If we see enough cases in which something is true and we never see a case in which it is false, we tend to conclude that it is always true.

> *I have seen 1000 black ravens.*
> *I have never seen a raven that is not black.*
> *Therefore, every raven is black.*
> Now try red Hondas.

Abduction is reasoning from effects to possible causes. Many things can cause an observed result. We often tend to infer a cause even when our enumeration of possible causes is incomplete.

> *If there is no fuel, the car will not start.*
> *If there is no spark, the car will not start.*
> *There is spark.*
> *The car will not start.*
> *Therefore, there is no fuel.*
> What if the car is in a vacuum chamber?

Reasoning by *analogy* is reasoning in which we infer a conclusion based on similarity of two situations, as in the following example.

> *The flow in a pipe is proportional to its diameter.*
> *Wires are like pipes.*
> *Therefore, the current in a wire is proportional to diameter.*
> Now try price.

Of all types of reasoning, deduction is the only one that *guarantees* its conclusions in all cases. It has some very special properties and holds a unique place in Logic. In this book, we concentrate entirely on deduction and leave these other forms of reasoning to others.

1.6 FORMALIZATION

So far, we have illustrated everything with sentences in English. While natural language works well in many circumstances, it is not without its problems. Natural language sentences can be complex; they can be ambiguous; and failing to understand the meaning of a sentence can lead to errors in reasoning.

Even very simple sentences can be troublesome. Here we see two grammatically legal sentences. They are the same in all but the last word, but their structure is entirely different. In the first, the main verb is *blossoms*, while in the second *blossoms* is a noun and the main verb is *sank*.

The cherry blossoms in the Spring.
The cherry blossoms in the Spring sank.

As another example of grammatical complexity, consider the following excerpt taken from the University of Michigan lease agreement. The sentence in this case is sufficiently long and the grammatical structure sufficiently complex that people must often read it several times to understand precisely what it says.

The University may terminate this lease when the Lessee, having made application and executed this lease in advance of enrollment, is not eligible to enroll or fails to enroll in the University or leaves the University at any time prior to the expiration of this lease, or for violation of any provisions of this lease, or for violation of any University regulation relative to resident Halls, or for health reasons, by providing the student with written notice of this termination 30 days prior to the effective date of termination, unless life, limb, or property would be jeopardized, the Lessee engages in the sales of purchase of controlled substances in violation of federal, state or local law, or the Lessee is no longer enrolled as a student, or the Lessee engages in the use or possession of firearms, explosives, inflammable liquids, fireworks, or other dangerous weapons within the building, or turns in a false alarm, in which cases a maximum of 24 hours notice would be sufficient.

As an example of ambiguity, suppose I were to write the sentence *There's a girl in the room with a telescope*. See Figure 1.6 for two possible meanings of this sentence. Am I saying that there is a girl in a room containing a telescope? Or am I saying that there is a girl in the room and she is holding a telescope?

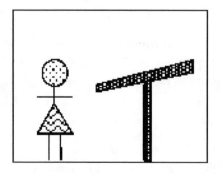 

Figure 1.6: *There's a girl in the room with a telescope.*

Such complexities and ambiguities can sometimes be humorous if they lead to interpretations the author did not intend. See the examples below for some infamous newspaper headlines with multiple interpretations. Using a formal language eliminates such unintentional ambiguities (and, for better or worse, avoids any unintentional humor as well).

Crowds Rushing to See Pope Trample 6 to Death
Journal Star, Peoria, 1980

Scientists Grow Frog Eyes and Ears   **British Left Waffles On Falkland Islands**
The Daily Camera, Boulder, 2000

Food Stamp Recipients Turn to Plastic   **Indian Ocean Talks**
The Miami Herald, 1991   The Plain Dealer, 1977

Fried Chicken Cooked in Microwave Wins Trip
The Oregonian, Portland, 1981

Figure 1.7: Various newspaper headlines.

As an illustration of errors that arise in reasoning with sentences in natural language, consider the following examples. In the first, we use the transitivity of the *better* relation to derive a conclusion about the relative quality of champagne and soda from the relative quality of champagne and beer and the relative quality or beer and soda. So far so good.

Champagne is better than beer.
Beer is better than soda.
Therefore, champagne is better than soda.

Now, consider what happens when we apply the same transitivity rule in the case illustrated below. The form of the argument is the same as before, but the conclusion is somewhat less believable. The problem in this case is that the use of *nothing* here is syntactically similar to the use of *beer* in the preceding example, but in English it means something entirely different.

Bad sex is better than nothing.
Nothing is better than good sex.
Therefore, bad sex is better than good sex.

Logic eliminates these difficulties through the use of a formal language for encoding information. Given the syntax and semantics of this formal language, we can give a precise definition for the notion of logical conclusion. Moreover, we can establish precise reasoning rules that produce all and only logical conclusions.

In this regard, there is a strong analogy between the methods of Formal Logic and those of high school algebra. To illustrate this analogy, consider the following algebra problem.

Xavier is three times as old as Yolanda. Xavier's age and Yolanda's age add up to twelve. How old are Xavier and Yolanda?

Typically, the first step in solving such a problem is to express the information in the form of equations. If we let x represent the age of Xavier and y represent the age of Yolanda, we can capture the essential information of the problem as shown below.

$$x - 3y = 0$$
$$x + y = 12$$

Using the methods of algebra, we can then manipulate these expressions to solve the problem. First we subtract the second equation from the first.

$$
\begin{array}{r}
x - 3y = 0 \\
x + y = 12 \\
\hline
-4y = -12
\end{array}
$$

Next, we divide each side of the resulting equation by -4 to get a value for y. Then substituting back into one of the preceding equations, we get a value for x.

$$x = 9$$
$$y = 3$$

Now, consider the following logic problem.

If Mary loves Pat, then Mary loves Quincy. If it is Monday and raining, then Mary loves Pat or Quincy. If it is Monday and raining, does Mary love Quincy?

As with the algebra problem, the first step is formalization. Let p represent the possibility that Mary loves Pat; let q represent the possibility that Mary loves Quincy; let m represent the possibility that it is Monday; and let r represent the possibility that it is raining.

With these abbreviations, we can represent the essential information of this problem with the following logical sentences. The first says that p *implies* q, i.e., if Mary loves Pat, then Mary loves Quincy. The second says that m *and* r *implies* p *or* q, i.e., if it is Monday and raining, then Mary loves Pat or Mary loves Quincy.

$$p \quad\quad \Rightarrow \quad q$$
$$m \wedge r \quad \Rightarrow \quad p \vee q$$

As with Algebra, Formal Logic defines certain operations that we can use to manipulate expressions. The operation shown below is a variant of what is called *Propositional Resolution*. The expressions above the line are the premises of the rule, and the expression below is the conclusion.

$$\frac{\begin{array}{lcl} p_1 \wedge ... \wedge p_k & \Rightarrow & q_1 \vee .. \vee q_l \\ r_1 \wedge ... \wedge r_m & \Rightarrow & s_1 \vee ... \vee s_n \end{array}}{p_1 \wedge ... \wedge p_k \wedge r_1 \wedge ... \wedge r_m \quad \Rightarrow \quad q_1 \vee ... \vee q_l \vee s_1 \vee ... \vee s_n}$$

There are two elaborations of this operation. (1) If a proposition on the left hand side of one sentence is the same as a proposition on the right hand side of the other sentence, it is okay to drop the two symbols, with the proviso that *only one* such pair may be dropped. (2) If a constant is repeated on the same side of a single sentence, all but one of the occurrences can be deleted.

We can use this operation to solve the problem of Mary's love life. Looking at the two premises above, we notice that p occurs on the left-hand side of one sentence and the right-hand side of the other. Consequently, we can cancel the p and thereby derive the conclusion that, if is Monday and raining, then Mary loves Quincy or Mary loves Quincy.

$$\frac{\begin{array}{lcl} p & \Rightarrow & q \\ m \wedge r & \Rightarrow & p \vee q \end{array}}{m \wedge r \quad \Rightarrow \quad q \vee q}$$

Dropping the repeated symbol on the right hand side, we arrive at the conclusion that, if it is Monday and raining, then Mary loves Quincy.

$$\frac{m \wedge r \quad \Rightarrow \quad q \vee q}{m \wedge r \quad \Rightarrow \quad q}$$

This example is interesting in that it showcases our formal language for encoding logical information. As with algebra, we use symbols to represent relevant aspects of the world in question, and we use operators to connect these symbols in order to express information about the things those symbols represent.

The example also introduces one of the most important operations in Formal Logic, viz. Resolution (in this case a restricted form of Resolution). Resolution has the property of being *complete* for an important class of logic problems, i.e., it is the *only* operation necessary to solve any problem in the class.

1.7   AUTOMATION

The existence of a formal language for representing information and the existence of a corresponding set of mechanical manipulation rules together have an important consequence, viz. the possibility of *automated reasoning* using digital computers.

The idea is simple. We use our formal representation to encode the premises of a problem as data structures in a computer, and we program the computer to apply our mechanical rules in a systematic way. The rules are applied until the desired conclusion is attained or until it is determined that the desired conclusion cannot be attained. (Unfortunately, in some cases, this determination cannot be made; and the procedure never halts. Nevertheless, as discussed in later chapters, the idea is basically sound.)

Although the prospect of automated reasoning has achieved practical realization only in the last few decades, it is interesting to note that the concept itself is not new. In fact, the idea of building machines capable of logical reasoning has a long tradition.

One of the first individuals to give voice to this idea was Leibnitz. He conceived of "a universal algebra by which all knowledge, including moral and metaphysical truths, can some day be brought within a single deductive system." Having already perfected a mechanical calculator for arithmetic, he argued that, with this universal algebra, it would be possible to build a machine capable of rendering the consequences of such a system mechanically.

Boole gave substance to this dream in the 1800s with the invention of Boolean algebra and with the creation of a machine capable of computing accordingly.

The early twentieth century brought additional advances in Logic, notably the invention of the predicate calculus by Russell and Whitehead and the proof of the corresponding completeness and incompleteness theorems by Godel in the 1930s.

The advent of the digital computer in the 1940s gave increased attention to the prospects for automated reasoning. Research in artificial intelligence led to the development of efficient algorithms for logical reasoning, highlighted by Robinson's invention of resolution theorem proving in the 1960s.

Today, the prospect of automated reasoning has moved from the realm of possibility to that of practicality, with the creation of *logic technology* in the form of automated reasoning systems, such as Vampire, Prover9, the Prolog Technology Theorem Prover, Epilog, and others.

The emergence of this technology has led to the application of logic technology in a wide variety of areas. The following paragraphs outline some of these uses.

Mathematics. Automated reasoning programs can be used to check proofs and, in some cases, to produce proofs or portions of proofs.

Engineering. Engineers can use the language of Logic to write specifications for their products and to encode their designs. Automated reasoning tools can be used to simulate designs and in some cases validate that these designs meet their specification. Such tools can also be used to diagnose failures and to develop testing programs.

Database Systems. By conceptualizing database tables as sets of simple sentences, it is possible to use Logic in support of database systems. For example, the language of Logic can be used to define virtual views of data in terms of explicitly stored tables, and it can be used to encode constraints on databases. Automated reasoning techniques can be used to compute new tables, to detect problems, and to optimize queries.

Data Integration The language of Logic can be used to relate the vocabulary and structure of disparate data sources, and automated reasoning techniques can be used to integrate the data in these sources.

Logical Spreadsheets. Logical spreadsheets generalize traditional spreadsheets to include logical constraints as well as traditional arithmetic formulas. Examples of such constraints abound. For example, in scheduling applications, we might have timing constraints or restrictions on who can reserve which rooms. In the domain of travel reservations, we might have constraints on adults and infants. In academic program sheets, we might have constraints on how many courses of varying types that students must take.

Law and Business. The language of Logic can be used to encode regulations and business rules, and automated reasoning techniques can be used to analyze such regulations for inconsistency and overlap.

1.8 READING GUIDE

Although Logic is a single field of study, there is more than one logic in this field. In the three main units of this book, we look at three different types of logic, each more sophisticated than the one before.

Propositional Logic is the logic of propositions. Symbols in the language represent "conditions" in the world, and complex sentences in the language express interrelationships among these conditions. The primary operators are Boolean connectives, such as *and, or,* and *not.*

Relational Logic expands upon Propositional Logic by providing a means for explicitly talking about individual objects and their interrelationships (not just monolithic conditions). In order to do so, we expand our language to include object constants and relation constants, variables and quantifiers.

Herbrand Logic takes us one step further by providing a means for describing worlds with infinitely many objects. The resulting logic is much more powerful than Propositional Logic and Relational Logic. Unfortunately, as we shall see, many of the nice computational properties of the first two logics are lost as a result.

Despite their differences, there are many commonalities among these logics. In particular, in each case, there is a language with a formal syntax and a precise semantics; there is a notion of logical entailment; and there are legal rules for manipulating expressions in the language.

These similarities allow us to compare the logics and to gain an appreciation of the fundamental tradeoff between expressiveness and computational complexity. On the one hand, the introduction of additional linguistic complexity makes it possible to say things that cannot be said in more restricted languages. On the other hand, the introduction of additional linguistic flexibility has adverse effects on computability. As we proceed though the material, our attention will range from the completely computable case of Propositional Logic to a variant that is not at all computable.

One final comment. In the hopes of preventing difficulties, it is worth pointing out a potential source of confusion. This book exists in the *meta* world. It contains sentences about sentences; it contains proofs about proofs. In some places, we use similar mathematical symbology both for sentences *in* Logic and sentences *about* Logic. Wherever possible, we try to be clear about this distinction, but the potential for confusion remains. Unfortunately, this comes with the territory. We are using Logic to study Logic. It is our most powerful intellectual tool.

RECAP

Logic is the study of information encoded in the form of logical sentences. Each logical sentence divides the set of all possible world into two subsets—the set of worlds in which the sentence is true and the set of worlds in which the set of sentences is false. A set of premises *logically entails* a conclusion if and only if the conclusion is true in every world in which all of the premises are true. *Deduction* is a form of symbolic reasoning that produces conclusions that are logically entailed by premises (distinguishing it from other forms of reasoning, such as *induction*, *abduction*, and *analogical reasoning*). A *proof* is a sequence of simple, more-or-less obvious deductive steps that justifies a conclusion that may not be immediately obvious from given premises. In Logic, we usually encode logical information as sentences in formal languages; and we use rules of inference appropriate to these languages. Such formal representations and methods are useful for us to use ourselves. Moreover, they allow us to automate the process of deduction, though the computability of such implementations varies with the complexity of the sentences involved.

1.9   EXERCISES

1.1.   Consider the state of the Sorority World depicted below.

	Abby	Bess	Cody	Dana
Abby		✓	✓	
Bess			✓	
Cody	✓	✓		✓
Dana		✓	✓	

For each of the following sentences, say whether or not it is true in this state of the world.

(*a*) *Abby likes Dana.*
(*b*) *Dana does not like Abby.*
(*c*) *Abby likes Cody or Dana.*
(*d*) *Abby likes someone who likes her.*
(*e*) *Somebody likes everybody.*

1.2.   Come up with a table of likes and dislikes for the Sorority World that makes *all* of the following sentences true. Note that there is more than one such table.

Dana likes Cody.
Abby does not like Dana.
Bess likes Cody or Dana.
Abby likes everyone whom Bess likes.
Cody likes everyone who likes her.
Nobody likes herself.

1.3.   Consider a set of Sorority World premises that are true in the four states of Sorority World shown in Figure 1.4. For each of the following sentences, say whether or not it is logically entailed by these premises.

(*a*) *Abby likes Bess or Bess likes Abby.*
(*b*) *Somebody likes herself.*
(*c*) *Everybody likes somebody.*

1.4.   Say whether or not the following reasoning patterns are logically correct.

(*a*) *All x are z. All y are z. Therefore, some x are y.*
(*b*) *Some x are y. All y are z. Therefore, some x are z.*
(*c*) *All x are y. Some y are z. Therefore, some x are z.*

CHAPTER 2

Propositional Logic

2.1 INTRODUCTION

Propositional Logic is concerned with propositions and their interrelationships. The notion of a proposition here cannot be defined precisely. Roughly speaking, a *proposition* is a possible condition of the world that is either true or false, e.g., the possibility that it is raining, the possibility that it is cloudy, and so forth. The condition need not be true in order for it to be a proposition. In fact, we might want to say that it is false or that it is true if some other proposition is true.

In this chapter, we first look at the syntactic rules that define the language of Propositional Logic. We then introduce the notion of a truth assignment and use it to define the meaning of Propositional Logic sentences. After that, we present a mechanical method for evaluating sentences for a given truth assignment, and we present a mechanical method for finding truth assignments that satisfy sentences. We conclude with some examples of Propositional Logic in formalizing Natural Language and Digital Circuits.

2.2 SYNTAX

In Propositional Logic, there are two types of sentences—simple sentences and compound sentences. Simple sentences express simple facts about the world. Compound sentences express logical relationships between the simpler sentences of which they are composed.

Simple sentences in Propositional Logic are often called *proposition constants* or, sometimes, *logical constants*. In what follows, we write proposition constants as strings of letters, digits, and underscores ("_"), where the first character is a lower case letter. For example, *raining* is a proposition constant, as are *rAiNiNg*, *r32aining*, and *raining_or_snowing*. *Raining* is not a proposition constant because it begins with an upper case character. 324567 fails because it begins with a number. *raining-or-snowing* fails because it contains hyphens (instead of underscores).

Compound sentences are formed from simpler sentences and express relationships among the constituent sentences. There are five types of compound sentences, viz. negations, conjunctions, disjunctions, implications, and biconditionals.

A *negation* consists of the negation operator ¬ and an arbitrary sentence, called the *target*. For example, given the sentence p, we can form the negation of p as shown below.

$$(\neg p)$$

A *conjunction* is a sequence of sentences separated by occurrences of the $\wedge$ operator and enclosed in parentheses, as shown below. The constituent sentences are called *conjuncts*. For example, we can form the conjunction of p and q as follows.

$$(p \wedge q)$$

A *disjunction* is a sequence of sentences separated by occurrences of the $\vee$ operator and enclosed in parentheses. The constituent sentences are called *disjuncts*. For example, we can form the disjunction of p and q as follows.

$$(p \vee q)$$

An *implication* consists of a pair of sentences separated by the $\Rightarrow$ operator and enclosed in parentheses. The sentence to the left of the operator is called the *antecedent*, and the sentence to the right is called the *consequent*. The implication of p and q is shown below.

$$(p \Rightarrow q)$$

A *biconditional* is a combination of an implication and a reverse implication. For example, we can express the biconditional of p and q as shown below.

$$(p \Leftrightarrow q)$$

Note that the constituent sentences within any compound sentence can be either simple sentences or compound sentences or a mixture of the two. For example, the following is a legal compound sentence.

$$((p \vee q) \Rightarrow r)$$

One disadvantage of our notation, as written, is that the parentheses tend to build up and need to be matched correctly. It would be nice if we could dispense with parentheses, e.g., simplifying the preceding sentence to the one shown below.

$$p \vee q \Rightarrow r$$

Unfortunately, we cannot do without parentheses entirely, since then we would be unable to render certain sentences unambiguously. For example, the sentence shown above could have resulted from dropping parentheses from either of the following sentences.

$$((p \vee q) \Rightarrow r)$$
$$(p \vee (q \Rightarrow r))$$

The solution to this problem is the use of *operator precedence*. The following table gives a hierarchy of precedences for our operators. The $\neg$ operator has higher precedence than $\wedge$; $\wedge$ has higher precedence than $\vee$; and $\vee$ has higher precedence than $\Rightarrow$ and $\Leftrightarrow$.

In sentences without parentheses, it is often the case that an expression is flanked by operators, one on either side. In interpreting such sentences, the question is whether the expression associates with the operator on its left or the one on its right. We can use precedence to make this determination. In particular, we agree that an operand in such a situation always associates with the operator of higher precedence. When an operand is surrounded by operators of equal precedence, the operand associates to the right. The following examples show how these rules work in various cases. The expressions on the right are the fully parenthesized versions of the expressions on the left.

$$\neg\, p \wedge q \qquad ((\neg\, p) \wedge q)$$
$$p \wedge \neg q \qquad (p \wedge (\neg\, q))$$
$$p \wedge q \vee r \qquad ((p \wedge q) \vee r)$$
$$p \vee q \wedge r \qquad (p \vee (q \wedge r))$$
$$p \Rightarrow q \Rightarrow r \qquad (p \Rightarrow (q \Rightarrow r))$$
$$p \Rightarrow q \Leftrightarrow r \qquad (p \Rightarrow (q \Leftrightarrow r))$$

Note that just because precedence allows us to delete parentheses in some cases does not mean that we can dispense with parentheses entirely. Consider the example shown earlier. Precedence eliminates the ambiguity by dictating that the sentence without parentheses is an implication with a disjunction as antecedent. However, this makes for a problem for those cases when we want to express a disjunction with an implication as a disjunct. In such cases, we must retain at least one pair of parentheses.

We end the section with two simple definitions that are useful in discussing Propositional Logic. A *propositional vocabulary* is a set of proposition constants. A *propositional language* is the set of all propositional sentences that can be formed from a propositional vocabulary.

2.3 SEMANTICS

The treatment of semantics in Logic is similar to its treatment in Algebra. Algebra is unconcerned with the real-world significance of variables. What is interesting are the relationships among the values of the variables expressed in the equations we write. Algebraic methods are designed to respect these relationships, independent of what the variables represent.

In a similar way, Logic is unconcerned with the real world significance of proposition constants. What is interesting is the relationship among the truth values of simple sentences and the truth values of compound sentences within which the simple sentences are contained. As with Algebra, logical reasoning methods are independent of the significance of proposition constants; all that matter is the form of sentences.

Although the values assigned to proposition constants are not crucial in the sense just described, in talking about Logic, it is sometimes useful to make truth assignments explicit and to consider various assignments or all assignments and so forth. Such an assignment is called a truth assignment.

Formally, a *truth assignment* for a propositional vocabulary is a function assigning a truth value to each of the proposition constants of the vocabulary. In what follows, we use the digit 1 as a synonym for *true* and 0 as a synonym for *false*; and we refer to the value of a constant or expression under a truth assignment i by superscripting the constant or expression with i as the superscript.

The assignment shown below is an example for the case of a propositional vocabulary with just three proposition constants, viz. p, q, and r.

$$p^i = 1$$
$$q^i = 0$$
$$r^i = 1$$

The following assignment is another truth assignment for the same vocabulary.

$$p^i = 0$$
$$q^i = 0$$
$$r^i = 1$$

Note that the formulas above are not themselves sentences in Propositional Logic. Propositional Logic does not allow superscripts and does not use the = symbol. Rather, these are informal, metalevel statements *about* particular truth assignments. Although talking about Propositional Logic using a notation similar to that of Propositional Logic can sometimes be confusing, it allows us to convey meta-information precisely and efficiently. To minimize problems, in this book we use such meta-notation infrequently and only when there is little chance of confusion.

Looking at the preceding truth assignments, it is important to bear in mind that, as far as logic is concerned, any truth assignment is as good as any other. Logic itself does not fix the truth assignment of individual proposition constants.

On the other hand, *given* a truth assignment for the proposition constants of a language, logic *does* fix the truth assignment for all compound sentences in that language. In fact, it is possible to determine the truth value of a compound sentence by repeatedly applying the following rules.

If the truth value of a sentence is *true*, the truth value of its negation is *false*. If the truth value of a sentence is *false*, the truth value of its negation is *true*.

φ	$\neg\varphi$
1	0
0	1

The truth value of a conjunction is *true* if and only if the truth values of its conjuncts are both *true*; otherwise, the truth value is *false*.

φ	ψ	$\varphi \wedge \psi$
1	1	1
1	0	0
0	1	0
0	0	0

The truth value of a disjunction is *true* if and only if the truth value of at least one its disjuncts is *true*; otherwise, the truth value is *false*. Note that this is the *inclusive or* interpretation of the $\vee$ operator and is differentiated from the *exclusive or* interpretation in which a disjunction is true if and only if an odd number of its disjuncts are true.

φ	ψ	$\varphi \vee \psi$
1	1	1
1	0	1
0	1	1
0	0	0

The truth value of an implication is *false* if and only if its antecedent is *true* and its consequent is *false*; otherwise, the truth value is *true*. This is called *material implication*.

φ	ψ	$\varphi \Rightarrow \psi$
1	1	1
1	0	0
0	1	1
0	0	1

A biconditional is *true* if and only if the truth values of its constituents agree, i.e., they are either both *true* or both *false*.

φ	ψ	$\varphi \Leftrightarrow \psi$
1	1	1
1	0	0
0	1	0
0	0	1

Using these definitions, it is easy to determine the truth values of compound sentences with proposition constants as constituents. As we shall see in the next section, we can determine the truth values of compound sentences with other compound sentences as parts by first evaluating the constituent sentences and then applying these definitions to the results.

We finish up this section with a few definitions for future use. We say that a truth assignment *satisfies* a sentence if and only if the sentence is *true* under that truth assignment. We say that a truth assignment *falsifies* a sentence if and only if the sentence is *false* under that truth assignment. A truth assignment satisfies a *set* of sentences if and only if it satisfies *every* sentence in the set. A truth assignment falsifies a *set* of sentences if and only if it falsifies *at least* one sentence in the set.

2.4   EVALUATION

Evaluation is the process of determining the truth values of compound sentences given a truth assignment for the truth values of proposition constants.

As it turns out, there is a simple technique for evaluating complex sentences. We substitute true and false values for the proposition constants in our sentence, forming an expression with 1s and 0s and logical operators. We use our operator semantics to evaluate subexpressions with these truth values as arguments. We then repeat, working from the inside out, until we have a truth value for the sentence as a whole.

As an example, consider the truth assignment i shown below.

$$p^i = 1$$
$$q^i = 0$$
$$r^i = 1$$

Using our evaluation method, we can see that i satisfies $(p \vee q) \wedge (\neg q \vee r)$.

$$(p \vee q) \wedge (\neg q \vee r)$$
$$(1 \vee 0) \wedge (\neg 0 \vee 1)$$
$$1 \wedge (\neg 0 \vee 1)$$
$$1 \wedge (1 \vee 1)$$
$$1 \wedge 1$$
$$1$$

Now consider truth assignment j defined as follows.

$$p^j = 0$$
$$q^j = 1$$
$$r^j = 0$$

In this case, j does not satisfy $(p \vee q) \wedge (\neg q \vee r)$.

$$(p \vee q) \wedge (\neg q \vee r)$$
$$(0 \vee 1) \wedge (\neg 1 \vee 0)$$
$$1 \wedge (\neg 1 \vee 0)$$
$$1 \wedge (0 \vee 0)$$
$$1 \wedge 0$$
$$0$$

Using this technique, we can evaluate the truth of arbitrary sentences in our language. The cost is proportional to the size of the sentence. Of course, in some cases, it is possible to economize and do even better. For example, when evaluating a conjunction, if we discover that the first conjunct is false, then there is no need to evaluate the second conjunct since the sentence as a whole must be false.

2.5   SATISFACTION

Satisfaction is the opposite of evaluation. We begin with one or more compound sentences and try to figure out which truth assignments satisfy those sentences. One nice feature of Propositional Logic is that there are effective procedures for finding truth assignments that satisfy Propositional Logic sentences. In this section, we look at a method based on truth tables.

A *truth table* for a propositional language is a table showing all of the possible truth assignments for the proposition constants in the language. The columns of the table correspond to the proposition constants of the language, and the rows correspond to different truth assignments for those constants.

The following figure shows a truth table for a propositional language with just three proposition constants (p, q, and r). Each column corresponds to one proposition constant, and each row corresponds to a single truth assignment. The truth assignments i and j defined in the preceding section correspond to the third and sixth rows of this table, respectively.

p	q	r
1	1	1
1	1	0
1	0	1
1	0	0
0	1	1
0	1	0
0	0	1
0	0	0

Note that, for a propositional language with n proposition constants, there are n columns in the truth table and 2^n rows.

In solving satisfaction problems, we start with a truth table for the proposition constants of our language. We then process our sentences in turn, for each sentence placing an x next to rows in the truth table corresponding to truth assignments that do not satisfy the sentence. The truth assignments remaining at the end of this process are all possible truth assignments of the input sentences.

As an example, consider the sentence $p \vee q \Rightarrow q \wedge r$. We can find all truth assignments that satisfy this sentence by constructing a truth table for p, q, and r. See below. We place an x next to each row that does not satisfy the sentence (rows 2, 3, 4, 6). Finally, we take the remaining rows (1, 5, 7, 8) as answers.

	p	q	r	
	1	1	1	
x	1	1	0	x
x	1	0	1	x
x	1	0	0	x
	0	1	1	
x	0	1	0	x
	0	0	1	
	0	0	0	

The disadvantage of the truth table method is computational complexity. As mentioned above, the size of a truth table for a language grows exponentially with the number of proposition constants in the language. When the number of constants is small, the method works well. When the number is large, the method becomes impractical. Even for moderate sized problems, it can be tedious. Even for an application like Sorority World, where there are only 16 proposition constants, there are 65,536 truth assignments.

Over the years, researchers have proposed ways to improve the performance of truth table checking. However, the best approach to dealing with large vocabularies is to use symbolic manipulation (i.e., logical reasoning and proofs) in place of truth table checking. We discuss these methods in Chapters 4 and 5.

2.6 EXAMPLE–NATURAL LANGUAGE

As an exercise in working with Propositional Logic, let's look at the encoding of various English sentences as formal sentences in Propositional Logic. As we shall see, the structure of English sentences, along with various key words, such as *if* and *no*, determine how such sentences should be translated.

The following examples concern three properties of people, and we assign a different proposition constant to each of these properties. We use the constant c to mean that a person is cool. We use the constant f to mean that a person is funny. And we use the constant p to mean that a person is popular.

As our first example, consider the English sentence *If a person is cool or funny, then he is popular*. Translating this sentence into the language of Propositional Logic is straightforward. The use of the words *if* and *then* suggests an implication. The condition (*cool or funny*) is clearly a disjunction, and the conclusion (*popular*) is just a simple fact. Using the vocabulary from the last paragraph, this leads to the Propositional Logic sentence shown below.

$$c \vee f \Rightarrow p$$

Next, we have the sentence *A person is popular only if he is either cool or funny*. This is similar to the previous sentence, but the presence of the phrase *only if* suggests that the conditionality goes the other way. It is equivalent to the sentence *If a person is popular, then he is either cool or funny*. And this sentence can be translated directly into Propositional Logic as shown below.

$$p \Rightarrow c \vee f$$

A person is popular if and only if he is either cool or funny. The use of the phrase *if and only if* suggests a biconditional, as in the translation shown below. Note that this is the equivalent to the conjunction of the two implications shown above. The biconditional captures this conjunction in a more compact form.

$$p \Leftrightarrow c \vee f$$

Finally, we have a negative sentence. *There is no one who is both cool and funny*. The word *no* here suggests a negation. To make it easier to translate into Propositional Logic, we can first rephrase this as *It is not the case that there is a person who is both cool and funny*. This leads directly to the following encoding.

$$\neg(c \wedge f)$$

Note that, just because we can translate sentences into the language of Propositional Logic does not mean that they are true. The good news is that we can use our evaluation procedure to determine which sentences are true and which are false?

Suppose we were to imagine a person who is cool and funny and popular, i.e., the proposition constants c and f and p are all true. Which of our sentences are true and which are false.

Using the evaluation procedure described earlier, we can see that, for this person, the first sentence is true.

$$c \vee f \Rightarrow p$$
$$(1 \vee 1) \Rightarrow 1$$
$$1 \Rightarrow 1$$
$$1$$

The second sentence is also true.

$$p \Rightarrow c \vee f$$
$$1 \Rightarrow (1 \vee 1)$$
$$1 \Rightarrow 1$$
$$1$$

Since the third sentence is really just the conjunction of the first two sentences, it is also true, which we can confirm directly as shown below.

$$p \Leftrightarrow c \vee f$$
$$1 \Leftrightarrow (1 \vee 1)$$
$$1 \Leftrightarrow 1$$
$$1$$

Unfortunately, the fourth sentence is not true, since the person in this case is both cool and funny.

$$\neg(c \wedge f)$$
$$\neg(1 \wedge 1)$$
$$\neg 1$$
$$0$$

In this particular case, three of the sentences are true, while one is false. The upshot of this is that there is no such person (assuming that the theory expressed in our sentences is correct). The good news is that there are cases where all four sentences are true, e.g. a person who is cool and popular but not funny or the case of a person who is funny and popular but not cool. Question to consider: What about a person is neither cool nor funny nor popular? Is this possible according to our theory? Which of the sentences would be true and which would be false?

2.7 EXAMPLE–DIGITAL CIRCUITS

Now let's consider the use of Propositional Logic in modeling a portion of the physical world, in this case, a digital circuit like the ones used in building computers.

The diagram below is a pictorial representation of such a circuit. There are three input *nodes*, some internal nodes, and two output nodes. There are five *gates* connecting these nodes to each other—two *xor* gates (the gates on the top), two *and* gates (the gates on the lower left), and one or gate (the gate on the lower right).

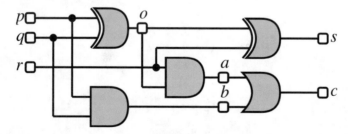

Figure 2.1: Click on p, q, r to toggle their values.

At a given point in time, a node in a circuit can be either *on* or *off*. The input nodes are set from outside the circuit. A gate sets its output either *on* or *off* based on the type of gate and the values of its input nodes. The output of an *and* gate is *on* if and only if both of its inputs are *on*. The value of an *or* node is *on* if and only if at least one of its inputs is *on*. The output of an *xor* gate is *on* if and only if its inputs disagree with each other.

Given the Boolean nature of signals on nodes and the deterministic character of gates, it is quite natural to model digital circuits in Propositional Logic. We can represent each node of a circuit as a proposition constant, with the idea that the node is *on* if and only if the constant is true. Using the language of Propositional Logic, we can capture the behavior of gates by writing sentences relating the values of the inputs nodes and out nodes of the gates.

The sentences shown below capture the five gates in the circuit shown above. Node o must be *on* if and only if nodes p and q disagree.

$$(p \wedge \neg q) \vee (\neg p \wedge q) \Leftrightarrow o$$
$$r \wedge o \Leftrightarrow a$$
$$p \wedge q \Leftrightarrow b$$
$$(o \wedge \neg r) \vee (\neg o \wedge r) \Leftrightarrow s$$
$$a \vee b \Leftrightarrow c$$

Once we have done this, we can use our formalization to analyze the circuit - to determine if it meets it specification, to test whether a particular instance is operating correctly, and to diagnose the problem in cases here it is not.

RECAP

The syntax of Propositional Logic begins with a set of *proposition constants*. Compound sentences are formed by combining simpler sentences with logical operators. In the version of Propositional Logic used here, there are five types of compound sentences—negations, conjunctions, disjunctions, implications, and biconditionals. A *truth assignment* for Propositional Logic is a mapping that assigns a truth value to each of the proposition constants in the language. A truth assignment *satisfies* a sentence if and only if the sentences is *true* under that truth assignment according to rules defining the logical operators of the language. *Evaluation* is the process of determining the truth values of a complex sentence, given a truth assignment for the truth values of proposition constants in that sentence. *Satisfaction* is the process of determining whether or not a sentence has a truth assignment that satisfies it.

2.8 EXERCISES

2.1. Say whether each of the following expressions is a syntactically legal sentence of Propositional Logic.

  (a) $p \wedge \neg p$
  (b) $\neg p \vee \neg p$
  (c) $\neg (q \vee r) \neg q \Rightarrow \neg \neg p$
  (d) $(p \wedge q) \vee (p \neg \wedge q)$
  (e) $p \vee \neg q \wedge \neg p \vee \neg q \Rightarrow p \vee q$

2.2. Consider a truth assignment in which p is true, q is false, r is true. Use this truth assignment to evaluate the following sentences.

(a) $p \Rightarrow q \wedge r$
(b) $p \Rightarrow q \vee r$
(c) $p \wedge q \Rightarrow r$
(d) $p \wedge q \Rightarrow \neg r$
(e) $p \wedge q \Leftrightarrow q \wedge r$

2.3.   A small company makes widgets in a variety of constituent materials (aluminum, copper, iron), colors (red, green, blue, grey), and finishes (matte, textured, coated). Although there are more than one thousand possible combinations of widget features, the company markets only a subset of the possible combinations. The following sentences are constraints that characterize the possibilities. Suppose that a customer places an order for a copper widget that is both green and blue with a matte coating. Your job is to determine which constraints are satisfied and which are violated.

(a)   *aluminum* $\vee$ *copper* $\vee$ *iron*
(b)   *aluminum* $\Rightarrow$ *grey*
(c)   *copper* $\wedge$ $\neg$*coated* $\Rightarrow$ *red*
(d)   *coated* $\wedge$ $\neg$*copper* $\Rightarrow$ *green*
(e)   *green* $\vee$ *blue* $\Leftrightarrow$ $\neg$*textured* $\wedge$ $\neg$*iron*

2.4.   Consider the sentences shown below. There are three proposition constants here, meaning that there are eight possible truth assignments. How many of these assignments satisfy all of these sentences?

$$p \vee q \vee r$$
$$p \Rightarrow q \wedge r$$
$$q \Rightarrow \neg r$$

2.5.   A small company makes widgets in a variety of constituent materials (aluminum, copper, iron), colors (red, green, blue, grey), and finishes (matte, textured, coated). Although there are more than one thousand possible combinations of widget features, the company markets only a subset of the possible combinations. The sentences below are some constraints that characterize the possibilities. Your job here is to select materials, colors, and finishes in such a way that *all* of the product constraints are satisfied. Note that there are multiple ways this can be done.

$$aluminum \vee copper \vee iron$$
$$red \vee green \vee blue \vee grey$$
$$aluminum \Rightarrow grey$$
$$copper \wedge \neg coated \Rightarrow red$$
$$iron \Rightarrow coated$$

2.6. Consider a propositional language with three proposition constants—*mushroom, purple*, and *poisonous*—each indicating the property suggested by its spelling. Using these proposition constants, encode the following English sentences as Propositional Logic sentences.

(*a*) *All purple mushrooms are poisonous.*

(*b*) *A mushroom is poisonous only if it is purple.*

(*c*) *A mushroom is not poisonous unless it is purple.*

(*d*) *No purple mushroom is poisonous.*

2.7. Consider the digital circuit described in Section 2.7. Suppose we set nodes p, q, and r to be *on*, and we observe that all of the other nodes are *on*. Running our evaluation procedure, we would see that the first sentence in our description of the circuit is not true. Hence the circuit is malfunctioning. Is there any combination of inputs p, q, and r that would result in all other nodes being *on* in a correctly functioning circuit? Hint: To answer this, you need consider a truth table with just eight rows (the possible values for nodes p, q, and r) since all other nodes are observed to be *on*.

CHAPTER 3

Logical Properties and Relationships

3.1 INTRODUCTION

Satisfaction is a relationship between specific sentences and specific truth assignments. In Logic, we are usually more interested in properties and relationships of sentences that hold across all truth assignments. We begin this chapter with a look at logical properties of individual sentences (as opposed to relationships among sentences)—validity, contingency, and unsatisfiability. We then look at three types of logical relationship between sentences—logical entailment, logical equivalence, and logical consistency. We conclude with a discussion of the connections between the logical properties of individual sentences and logical relationships between sentences.

3.2 LOGICAL PROPERTIES

In the preceding chapter, we saw that some sentences are true in some truth assignments and false in others. However, this is not always the case. There are sentences that are always true and sentences that are always false as well as sentences that are sometimes true and sometimes false. This leads to a partition of sentences into three disjoint categories.

A sentence is *valid* if and only if it is satisfied by *every* truth assignment. For example, the sentence $(p \lor \neg p)$ is valid. If a truth assignment makes p true, then the first disjunct is true and the disjunction as a whole true. If a truth assignment makes p false, then the second disjunct is true and the disjunction as a whole is true.

A sentence is *unsatisfiable* if and only if it is not satisfied by any truth assignment. For example, the sentence $(p \land \neg p)$ is unsatisfiable. No matter what truth assignment we take, the sentence is always false. The argument is analogous to the argument in the preceding paragraph.

Finally, a sentence is *contingent* if and only if there is some truth assignment that satisfies it and some truth assignment that falsifies it. For example, the sentence $(p \land q)$ is contingent. If p and q are both true, it is true. If p and q are both false, it is false.

In one sense, valid sentences and unsatisfiable sentences are useless. Valid sentences do not rule out any possible truth assignments, and unsatisfiable sentences rule out all truth assignments. Thus, they tell us nothing about the world. In this regard, contingent sentences are the most useful. On the other hand, from a logical perspective, valid and unsatisfiable sentences are useful in that, as we shall see, they serve as the basis for legal transformations that we can perform on other logical sentences.

For many purposes, it is useful to group validity, contingency, and unsatisfiability into two groups. We say that a sentence is *satisfiable* if and only if it is valid or contingent. In other words the sentence is satisfied by at least one truth assignment. We say that a sentence is *falsifiable* if and only if it is unsatisfiable or contingent. In other words, the sentence is falsified by at least one truth assignment.

3.3 LOGICAL EQUIVALENCE

Intuitively, we think of two sentences as being equivalent if they say the same thing, i.e., they are true in exactly the same worlds. More formally, we say that a sentence φ is *logically equivalent* to a sentence ψ if and only if every truth assignment that satisfies φ satisfies ψ *and* every truth assignment that satisfies ψ satisfies φ.

The sentence $\neg(p \lor q)$ is logically equivalent to the sentence $(\neg p \land \neg q)$. If p and q are both true, then both sentences are false. If either p is true or q is true, then the disjunction in the first sentence is true and the sentence as a whole false. Similarly, either p is true or q is true, then one of the conjuncts in the second sentence is false and so the sentence as a whole is false. Since both sentences are satisfied by the same truth assignments, they are logically equivalent.

By contrast, the sentences $(p \land q)$ and $(p \lor q)$ are not logically equivalent. The first is false when p is true and q is false, while in this situation the disjunction is true. Hence, they are not logically equivalent.

One way of determining whether or not two sentences are logically equivalent is to check the truth table for the proposition constants in the language. This is called the truth table method. (1) First, we form a truth table for the proposition constants and add a column for each of the sentences. (2) We then evaluate the two expressions. (3) Finally, we compare the results. If the values for the two sentences are true in every case, then the two sentences are logically equivalent; otherwise, they are not.

As an example, let's use this method to show that $\neg(p \lor q)$ is logically equivalent to $(\neg p \land \neg q)$. We set up our truth table, add a column for each of our two sentences, and evaluate them for each truth assignment. Having done so, we notice that every row that satisfies the first sentence also satisfies the second. Hence, the sentences are logically equivalent.

p	q	$\neg(p \lor q)$	$\neg p \land \neg q$
1	1	0	0
1	0	0	0
0	1	0	0
0	0	1	1

Now, let's do the same for $(p \wedge q)$ and $(p \vee q)$. We set up our table as before and evaluate our sentences. In this case, there is only one row that satisfies first sentence while three rows satisfy the second. Consequently, they are not logically equivalent.

p	q	$p \wedge q$	$p \vee q$
1	1	1	1
1	0	0	1
0	1	0	1
0	0	0	0

One of the interesting properties of logically equivalence is substitutability. If a sentence φ is logically equivalent to a sentence ψ, then we can substitute φ for ψ in any Propositional Logic sentence and the result will be logically equivalent to the original sentence. (Note that this is not quite true in Relational Logic, as we shall see when we cover that logic.)

3.4   LOGICAL ENTAILMENT

We say that a sentence φ *logically entails* a sentence ψ (written $\varphi \models \psi$) if and only if every truth assignment that satisfies φ also satisfies ψ. More generally, we say that a set of sentences Δ *logically entails* a sentence ψ (written $\Delta \models \psi$) if and only if every truth assignment that satisfies all of the sentences in Δ also satisfies ψ.

For example, the sentence p logically entails the sentence $(p \vee q)$. Since a disjunction is true whenever one of its disjuncts is true, then $(p \vee q)$ must be true whenever p is true. On the other hand, the sentence p does *not* logically entail $(p \wedge q)$. A conjunction is true if and only if *both* of its conjuncts are true, and q may be false. Of course, any set of sentences containing both p and q does logically entail $(p \wedge q)$.

Note that the relationship of logical entailment is a purely logical one. Even if the premises of a problem do not logically entail the conclusion, this does not mean that the conclusion is necessarily false, even if the premises are true. It just means that it is *possible* that the conclusion is false.

Once again, consider the case of $(p \wedge q)$. Although p does not logically entail this sentence, it is *possible* that both p and q are true and, therefore, $(p \wedge q)$ is true. However, the logical entailment does not hold because it is also possible that q is false and, therefore, $(p \wedge q)$ is false.

Note also that logical entailment is not the same as logical equivalence. The sentence p logically entails $(p \vee q)$, but $(p \vee q)$ does not logically entail p. Logical entailment is not analogous to arithmetic equality; it is closer to arithmetic inequality.

As with logical equivalence, we can use truth tables to determine whether or not a set of premises logically entails a possible conclusion by checking the truth table for the proposition constants in the language. (1) We form a truth table for the proposition constants and add a column for the premises and a column for the conclusion. (2) We then evaluate the premises.

(3) We evaluate the conclusion. (4) Finally, we compare the results. If every row that satisfies the premises also satisfies the conclusion, then the premises logically entail the conclusion.

As an example, let's use this method to show that p logically entails $(p \vee q)$. We set up our truth table and add a column for our premise and a column for our conclusion. In this case the premise is just p and so evaluation is straightforward; we just copy the column. The conclusion is true if and only if p is true or q is true. Finally, we notice that every row that satisfies the premise also satisfies the conclusion.

p	q	p	$p \vee q$
1	1	1	1
1	0	1	1
0	1	0	1
0	0	0	0

Now, let's do the same for the premise p and the conclusion $(p \wedge q)$. We set up our table as before and evaluate our premise. In this case, there is only one row that satisfies our conclusion. Finally, we notice that the assignment in the second row satisfies our premise but does not satisfy our conclusion; so logical entailment does not hold.

p	q	p	$p \wedge q$
1	1	1	1
1	0	1	0
0	1	0	0
0	0	0	0

Now, let's look at the problem of determining whether the set of propositions $\{p, q\}$ logically entails $(p \wedge q)$. Here we set up our table as before, but this time we have two premises to satisfy. Only one truth assignment satisfies both premises, and this truth assignment also satisfies the conclusion; hence in this case logical entailment does hold.

p	q	p	q	$p \wedge q$
1	1	1	1	1
1	0	1	0	0
0	1	0	1	0
0	0	0	0	0

As a final example, let's return to the love life of the fickle Mary. Here is the problem from the course introduction. We know $(p \Rightarrow q)$, i.e., if Mary loves Pat, then Mary loves Quincy. We know $(m \Rightarrow p \vee q)$, i.e., if it is Monday, then Mary loves Pat or Quincy. Let's confirm that, if it is Monday, then Mary loves Quincy. We set up our table and evaluate our premises and our conclusion. Both premises are satisfied by the truth assignments on rows 1, 3, 5, 7, and 8; and we notice that those truth assignments make the conclusion true. Hence, the logical entailment holds.

m	p	q	$m \Rightarrow p \vee q$	$p \Rightarrow q$	$m \Rightarrow q$
1	1	1	1	1	1
1	1	0	1	0	0
1	0	1	1	1	1
1	0	0	0	1	0
0	1	1	1	1	1
0	1	0	1	0	1
0	0	1	1	1	1
0	0	0	1	1	1

3.5   LOGICAL CONSISTENCY

A sentence φ is *consistent with* a sentence ψ if and only if there is a truth assignment that satisfies both φ and ψ. A sentence ψ is *consistent with* a set of sentences Δ if and only if there is a truth assignment that satisfies both Δ and ψ.

For example, the sentence $(p \vee q)$ is consistent with the sentence $(p \wedge q)$. However, it is *not* consistent with $(\neg p \wedge \neg q)$.

As with logical equivalence and logical entailment, we can use the truth table method to determine logical consistency. The following truth table shows all truth assignments for the propositional constants in the examples just mentioned. The third column shows the truth values for the first sentence; the fourth column shows the truth values for the second sentence, and the fifth column shows the truth values for the third sentence. The second and third truth assignments here make $(p \vee q)$ true and also $(\neg p \vee \neg q)$; hence $(p \vee q)$ and $(\neg p \vee \neg q)$ are consistent. By contrast, none of the truth assignments that makes $(p \vee q)$ true makes $(\neg p \wedge \neg q)$ true; hence, they are not consistent.

p	q	$p \vee q$	$\neg p \vee \neg q$	$\neg p \wedge \neg q$
1	1	1	0	0
1	0	1	1	0
0	1	1	1	0
0	0	0	1	1

The distinction between entailment and consistency is a subtle one and deserves some attention. Just because two sentences are consistent does not mean that they are logically equivalent or that either sentence logically entails the other.

Consider the sentences in the previous example. As we have seen, the first sentence and the second sentence are logically consistent, but they are clearly not logically equivalent and neither sentence logically entails the other.

Conversely, if one sentence logically entails another this does not necessarily mean that the sentences are consistent. This situation occurs when one of the sentences is unsatisfiable. If a sentence is unsatisfiable, there are no truth assignments that satisfy it. So, by definition, every truth assignment that satisfies the sentence (there are none) trivially satisfies the other sentence.

An interesting consequence of this fact is that any unsatisfiable sentence or set of sentences logically entails *everything*. Weird fact, but it follows directly from our definitions. And it makes clear why we want to avoid unsatisfiable sets of sentences in logical reasoning.

3.6   CONNECTIONS BETWEEN PROPERTIES AND RELATIONSHIPS

Before we end this chapter, it is worth noting that there are some strong connections between logical properties like validity and satisfiability and the logical relationships introduced in the preceding three sections.

First of all, there is a connection between the logical equivalence of two sentences and the validity of the biconditional sentence built from the two sentences. In particular, we have the following theorem expressing this connection.

Equivalence Theorem: A sentence φ and a sentence ψ are logically equivalent if and only if the sentence $(\varphi \Leftrightarrow \psi)$ is valid.

Why is this true? Consider the definition of logical equivalence. Two sentences are logically equivalent if and only if they are satisfied by the same set of truth assignments. Now recall the semantics of sentences involving the biconditional operator. A biconditional is true if and only if the truth values of the conditional sentences are the same. Clearly, if two sentences are logically equivalent, they are satisfied by the same truth assignments, and so the corresponding biconditional must be valid. Conversely, if a biconditional is valid, the two component sentences must be satisfied by the same truth assignments and so they are logically equivalent.

There is a similar connection between logical entailment between two sentences and the validity of the corresponding implication. And there is a natural extension to cases of logical entailment involving finite sets of sentences. The following theorem summarizes these results.

Deduction Theorem: A sentence φ logically entails a sentence ψ if and only if $(\varphi \Rightarrow \psi)$ is valid. More generally, a finite set of sentences $\{\varphi_1, \dots, \varphi_n\}$ logically entails φ if and only if the compound sentence $(\varphi_1 \wedge \dots \wedge \varphi_n \Rightarrow \varphi)$ is valid.

If a sentence φ logically entails a sentence ψ, it means that any truth assignment that satisfies φ also satisfies ψ. Looking at the semantics of implications, we see that an implication is true if and only if every truth assignment that makes the antecedent true also makes the consequent true. Consequently, logical entailment holds exactly when the corresponding implication is valid.

There is also a connection between logical entailment and unsatisfiability. In particular, if a set Δ of sentences logically entails a sentence φ, then Δ together with the negation of φ must be unsatisfiable. The reverse is also true.

Unsatisfiability Theorem: A set Δ of sentences logically entails a sentence φ if and only if the set of sentences $\Delta \cup \{\neg\varphi\}$ is unsatisfiable.

Suppose that Δ logically entails φ. If a truth assignment satisfies Δ, then it must also satisfy φ. But then it cannot satisfy $\neg\varphi$. Therefore, $\Delta \cup \{\neg\varphi\}$ is unsatisfiable. Suppose that $\Delta \cup \{\neg\varphi\}$ is unsatisfiable. Then every truth assignment that satisfies Δ must fail to satisfy $\neg\varphi$, i.e., it must satisfy φ. Therefore, Δ must logically entail φ.

An interesting consequence of this result is that we can determine logical entailment by checking for unsatisfiability. This turns out to be useful in various logical proof methods, as described in the following chapters.

Finally, consider the definition of logical consistency. A sentence φ is logically consistent with a sentence ψ if and only if there is a truth assignment that satisfies both φ and ψ. This is equivalent to saying that the sentence $(\varphi \wedge \psi)$ is satisfiable.

Consistency Theorem: A sentence φ is logically consistent with a sentence ψ if and only if the sentence $(\varphi \wedge \psi)$ is satisfiable. More generally, a sentence φ is logically consistent with a finite set of sentences $\{\varphi_1, \dots, \varphi_n\}$ if and only if the compound sentence $(\varphi_1 \wedge \dots \wedge \varphi_n \wedge \varphi)$ is satisfiable.

In thinking about these various connections, the main thing to keep in mind is that logical properties and logical relationships are metalevel. They are things we assert in talking *about* logical sentences; they are not sentences *within* our formal language. By contrast, implications and biconditionals and conjunctions are statements *within* our formal language; they are not metalevel statements. What the preceding paragraphs tell us is that we can *implicitly* express some logical relationships within our formal language by writing the corresponding biconditionals and implications and conjunctions and checking for the logical properties of these sentences.

RECAP

A sentence is *valid* if and only if it is satisfied by *every* truth assignment. A sentence is *unsatisfiable* if and only if it is not satisfied by any truth assignment. A sentence is *contingent* if and only if it is both satisfiable and falsifiable, i.e., it is neither valid nor unsatisfiable. A sentence is *satisfiable* if and only if it is either valid or contingent. A sentence is *falsifiable* if and only if it is unsatisfiable or contingent. A sentence φ is *logically equivalent* to a sentence ψ if and only if every truth assignment that satisfies φ satisfies ψ *and* every truth assignment that satisfies ψ satisfies φ. A set of sentences Δ *logically entails* a sentence φ (written $\Delta \models \varphi$) if and only if every truth assignment that satisfies Δ also satisfies φ. A sentence φ is *consistent with* a set of sentences Δ if and only if there is a truth assignment that satisfies both Δ and φ. The *Equivalence Theorem* states that sentence φ and a sentence ψ are logically equivalent if and only the sentence $(\varphi \Leftrightarrow \psi)$ is valid. The *Deduction Theorem* states that a sentence φ logically entails a sentence ψ if and only the sentence $(\varphi \Rightarrow \psi)$ is valid. More generally, a finite set of sentences $\{\varphi_1, \dots, \varphi_n\}$ logically entails φ if and only if the compound sentence $(\varphi_1 \wedge \dots \wedge \varphi_1 \Rightarrow \varphi)$ is valid. The *Unsatisfiability Theorem* states that a set Δ of sentences logically entails a sentence φ if and only if the set of sentences Δ

$\cup \{\neg\varphi\}$ is unsatisfiable. The *Consistency Theorem* states that a sentence φ is consistent with a set of sentences Δ if and only if the set of sentences $\Delta \cup \{\varphi\}$ is satisfiable. A sentence φ is consistent with a set of sentences $\{\varphi_1, \ldots, \varphi_n\}$ if and only if the compound sentence $(\varphi_1 \wedge \ldots \wedge \varphi_1 \wedge \varphi)$ is satisfiable. Finally, a consequence of our definitions—any unsatisfiable set of sentences logically entails everything.

3.7   EXERCISES

3.1.   Say whether each of the following sentences is valid, contingent, or unsatisfiable.

(a) $(p \Rightarrow q) \vee (q \Rightarrow p)$
(b) $p \wedge (p \Rightarrow \neg q) \wedge q$
(c) $(p \Rightarrow (q \wedge r)) \Leftrightarrow (p \Rightarrow q) \wedge (p \Rightarrow r)$
(d) $(p \Rightarrow (q \Rightarrow r)) \Rightarrow ((p \wedge q) \Rightarrow r)$
(e) $(p \Rightarrow q) \wedge (p \Rightarrow \neg q)$
(f) $(\neg p \vee \neg q) \Rightarrow \neg(p \wedge q)$
(g) $((\neg p \Rightarrow q) \Rightarrow (\neg q \Rightarrow p)) \wedge (p \vee q)$
(h) $(\neg p \vee q) \Rightarrow (q \wedge (p \Leftrightarrow q))$
(i) $((\neg r \Rightarrow \neg p \wedge \neg q) \vee s) \Leftrightarrow (p \vee q \Rightarrow r \vee s)$
(j) $(p \wedge (q \Rightarrow r)) \Leftrightarrow ((\neg p \vee q) \Rightarrow (p \wedge r))$

3.2.   For each of the following pairs of sentences, determine whether or not the sentences are logically equivalent.

(a) $(p \Rightarrow q \vee r)$ and $(p \wedge q \Rightarrow r)$
(b) $(p \Rightarrow (q \Rightarrow r))$ and $(p \wedge q \Rightarrow r)$
(c) $(p \wedge q \Rightarrow r)$ and $(p \wedge r \Rightarrow q)$
(d) $((p \Rightarrow q \vee r) \wedge (p \Rightarrow r))$ and $(q \Rightarrow r)$
(e) $((p \Rightarrow q) \vee (q \Rightarrow r))$ and $(p \vee \neg p)$

3.3.   Use the Truth Table Method to answer the following questions about logical entailment.

(a) $\{p \Rightarrow q \vee r\} \models (p \Rightarrow r)$
(b) $\{p \Rightarrow r\} \models (p \Rightarrow q \vee r)$
(c) $\{q \Rightarrow r\} \models (p \Rightarrow q \vee r)$
(d) $\{p \Rightarrow q \vee r, p \Rightarrow r\} \models (q \Rightarrow r)$
(e) $\{p \Rightarrow q \vee r, q \Rightarrow r\} \models (p \Rightarrow r)$

3.4.   Let Γ and Δ be sets of sentences in Propositional Logic, and let φ and ψ be individual sentences in Propositional Logic. State whether each of the following statements is true or false.

(a) If $\Gamma \models \varphi$ and $\Delta \models \varphi$, then $\Gamma \cap \Delta \models \varphi$.

(b) If $\Gamma \models \varphi$ and $\Delta \models \varphi$, then $\Gamma \cup \Delta \models \varphi$.

(c) If $\Gamma \models \varphi$ and $\Delta \,\#\, \varphi$, then $\Gamma \cup \Delta \models \varphi$.

(d) If $\Gamma \,\#\, \psi$, then $\Gamma \models \neg\psi$.

(e) If $\Gamma \models \neg\psi$, then $\Gamma \,\#\, \psi$.

3.5.   In each of the following cases, determine whether the given individual sentence is consistent with the given set of sentences.

(a) $\{p \vee q,\ p \vee \neg q,\ \neg p \vee q\}$ and $(\neg p \vee \neg q)$

(b) $\{p \Rightarrow r,\ q \Rightarrow r,\ p \vee q\}$ and r

(c) $\{p \Rightarrow r,\ q \Rightarrow r,\ p \vee q\}$ and $\neg r$

(d) $\{p \Rightarrow q \vee r,\ q \Rightarrow r\}$ and $p \wedge q$

(e) $\{p \Rightarrow q \vee r,\ q \Rightarrow r\}$ and $q \wedge r$

3.6.   Logical equivalence, logical entailment, and logical consistency are related to each other in interesting ways, but they are not identical. Answer the following true or false questions about the relationships between these concepts.

(a) If φ is equivalent to ψ, then φ entails ψ.

(b) If φ is equivalent to ψ, then φ is consistent with ψ.

(c) If φ entails ψ, then φ is equivalent to ψ.

(d) If φ entails ψ, then φ is consistent with ψ.

(e) If φ is consistent with ψ, then φ is equivalent to ψ.

(f) If φ is consistent with ψ, then φ entails ψ.

CHAPTER 4

Propositional Proofs

4.1 INTRODUCTION

Checking logical entailment with truth tables has the merit of being conceptually simple. However, it is not always the most practical method. The number of truth assignments of a language grows exponentially with the number of logical constants. When the number of logical constants in a propositional language is large, it may be impossible to process its truth table.

Proof methods provide an alternative way of checking logical entailment that addresses this problem. In many cases, it is possible to create a proof of a conclusion from a set of premises that is much smaller than the truth table for the language; moreover, it is often possible to find such proofs with less work than is necessary to check the entire truth table.

We begin this lesson with a discussion of linear reasoning and linear proofs. We then move on to hypothetical reasoning and structured proofs. Once we have seen both linear and structured proofs, we show how they are combined in the popular Fitch proof system, and we provide some tips for finding proofs using the Fitch system. We finish with definitions for soundness and completeness—the standards by which proof systems are judged.

4.2 LINEAR REASONING

As we saw in the introductory lesson, the essence of logical reasoning is symbolic manipulation. We start with premises, apply rules of inference to derive conclusions, stringing together such derivations to form logical proofs. The idea is simple. Getting the details right requires a little care. Let's start by defining schemas and rules of inference.

A *schema* is an expression satisfying the grammatical rules of our language except for the occurrence of *metavariables* (written here as Greek letters) in place of various subparts of the expression. For example, the following expression is a schema with metavariables φ and ψ.

$$\varphi \Rightarrow \psi$$

A *rule of inference* is a pattern of reasoning consisting of some schemas, called *premises*, and one or more additional schemas, called *conclusions*. Rules of inference are often written as shown below. The schemas above the line are the premises, and the schemas below the line are the conclusions.

$$\varphi \Rightarrow \psi$$
$$\underline{\varphi}$$
$$\psi$$

The rule in this case is called *Implication Elimination* (or IE), because it eliminates the implication from the first premise.

Implication Creation (IC), shown below, is another example. This rule tells us that, if a sentence ψ is true, we can infer $(\varphi \Rightarrow \psi)$ for any φ whatsoever.

$$\frac{\psi}{\varphi \Rightarrow \psi}$$

Implication Distribution (ID) tells us that implication can be distributed over other implications. If $(\varphi \Rightarrow (\psi \Rightarrow \chi))$ is true, then we can infer $((\varphi \Rightarrow \psi) \Rightarrow (\varphi \Rightarrow \chi))$.

$$\frac{\varphi \Rightarrow (\psi \Rightarrow \chi)}{(\varphi \Rightarrow \psi) \Rightarrow (\varphi \Rightarrow \chi)}$$

An instance of a rule of inference is the rule obtained by consistently substituting sentences for the metavariables in the rule. For example, the following is an instance of Implication Elimination.

$$\begin{array}{c} p \Rightarrow q \\ \underline{p} \\ q \end{array}$$

If a metavariable occurs more than once, the same expression must be used for every occurrence. For example, in the case of Implication Elimination, it would not be acceptable to replace one occurrence of φ with one expression and the other occurrence of φ with a different expression.

Note that the replacement can be an arbitrary expression so long as the result is a legal expression. For example, in the following instance of Implication Elimination, we have replaced the variables by compound sentences.

$$\begin{array}{c} (p \Rightarrow q) \Rightarrow (q \Rightarrow r) \\ \underline{(p \Rightarrow q)} \\ (q \Rightarrow r) \end{array}$$

Remember that there are infinitely many sentences in our language. Even though we start with finitely many propositional constants (in a propositional vocabulary) and finitely many operators, we can combine them in arbitrarily many ways. The upshot is that there are infinitely many instances of any rule of inference involving metavariables.

A rule *applies* to a set of sentences if and only if there is an instance of the rule in which all of the premises are in the set. In this case, the conclusions of the instance are the results of the rule application.

For example, if we had a set of sentences containing the sentence p and the sentence $(p \Rightarrow q)$, then we could apply Implication Elimination to derive q as a result. If we had a set of sentences containing the sentence $(p \Rightarrow q)$ and the sentence $(p \Rightarrow q) \Rightarrow (q \Rightarrow r)$, then we could apply Implication Elimination to derive $(q \Rightarrow r)$ as a result.

In using rules of inference, it is important to remember that they apply only to top-level sentences, not to components of sentences. While applying to components sometimes works, it can also lead to incorrect results.

As an example of such a problem, consider the incorrect application of Implication Elimination shown below. Suppose we believe $(p \Rightarrow q)$ and $(p \Rightarrow r)$. We might try to apply Implication Elimination here, taking the first premise as the implication and taking the occurrence of p in the second premise as the matching condition, leading us to conclude $(q \Rightarrow r)$.

$$\frac{\begin{array}{c} p \Rightarrow q \\ p \Rightarrow r \end{array}}{q \Rightarrow r}$$

Unfortunately, this is not a proper logical conclusion from the premises, as we all know from experience and as we can quickly determine by looking at the associated truth table. It is important to remember that rules of inference apply only to top-level sentences.

By writing down premises, writing instances of axiom schemas, and applying rules of inference, it is possible to derive conclusions that cannot be derived in a single step. This idea of stringing things together in this way leads to the notion of a linear proof.

A *linear proof* of a conclusion from a set of premises is a sequence of sentences terminating in the conclusion in which each item is either (1) a premise, (2) an instance of an axiom schema, or (3) the result of applying a rule of inference to earlier items in sequence.

Here is an example. Suppose we have the set of sentences we saw earlier. We start our proof by writing out our premises. We believe p; we believe $(p \Rightarrow q)$; and we believe that $(p \Rightarrow q) \Rightarrow (q \Rightarrow r)$. Using Implication Elimination on the first premise and the second premise, we derive q. Applying Implication Elimination to the second premise and the third premise, we derive $(q \Rightarrow r)$. Finally, we use the derived premises on lines 4 and 5 to arrive at our desired conclusion.

1.	p	Premise
2.	$p \Rightarrow q$	Premise
3.	$(p \Rightarrow q) \Rightarrow (q \Rightarrow r)$	Premise
4.	q	Implication Elimination: 2, 1
5.	$q \Rightarrow r$	Implication Elimination: 3, 2
6.	r	Implication Elimination: 5, 4

Here is another example. Whenever p is true, q is true. Whenever q is true, r is true. With these as premises, we can prove that, whenever p is true, r is true. On line 3, we use Implication Creation to derive $(p \Rightarrow (q \Rightarrow r))$. On line 4, we use Implication Distribution to distribute the implication in line 3. Finally, on line 5, we use Implication Elimination to produce the desired result.

1.	$p \Rightarrow q$	Premise
2.	$q \Rightarrow r$	Premise
3.	$p \Rightarrow (q \Rightarrow r)$	Implication Creation: 2
4.	$(p \Rightarrow q) \Rightarrow (p \Rightarrow r)$	Implication Distribution: 3
5.	$p \Rightarrow r$	Implication Elimination: 4, 1

Let R be a set of rules of inference. If there exists a proof of a sentence φ from a set Δ of premises using the rules of inference in R, we say that φ is *provable* from Δ using R. We usually write this as $\Delta \vdash_R \varphi$, using the provability operator $\vdash$ (which is sometimes called *single turnstile*). If the set of rules is clear from context, we usually drop the subscript, writing just $\Delta \vdash \varphi$.

Note that the set of rules presented here is not powerful enough to prove everything that is entailed by a set of premises in Propositional Logic. There is no support for using or deducing negations or conjunctions or disjunctions or biconditionals. Even if we restrict ourselves to implications, we need more rules. While such rules of inference exist, they are a little complicated. For many people, it is easier to reason about implications using hypothetical reasoning.

4.3   HYPOTHETICAL REASONING

Structured proofs are similar to linear proofs in that they are sequences of reasoning steps. However, they differ from linear proofs in that they have more structure. In particular, sentences can be grouped into subproofs nested within outer superproofs.

As an example, consider the structured proof shown below. It resembles a linear proof except that we have grouped the sentences on lines 3 through 5 into a subproof within our overall proof.

1.	$p \Rightarrow q$	Premise
2.	$q \Rightarrow r$	Premise
3.	p	Assumption
4.	q	Implication Elimination: 3, 1
5.	r	Implication Elimination: 4, 2
6.	$p \Rightarrow r$	Implication Introduction: 3, 5

The main benefit of structured proofs is that they allow us to prove things that cannot be proved using only ordinary rules of inference. In structured proofs, we can make assumptions within subproofs; we can prove conclusions from those assumptions; and, from those derivations, we can derive implications outside of those subproofs, with our assumptions as antecedents and our conclusions as consequents.

The structured proof above illustrates this. On line 3, we begin a subproof with the assumption that p is true. Note that p is not a premise in the overall problem. In a subproof, we can make whatever assumptions that we like. From p, we derive q using the premise on line 1; and, from that q, we prove r using the premise on line 2. That terminates the subproof. Finally, from this subproof, we derive $(p \Rightarrow r)$ in the outer proof. Given p, we can prove r; and so we know $(p \Rightarrow r)$. The rule used in this case is called Implication Introduction, or II for short.

As this example illustrates, there are three basic operations involved in creating useful subproofs—(1) making assumptions, (2) using ordinary rules of inference to derive conclusions, and (3) using structured rules of inference to derive conclusions outside of subproofs. Let's look at each of these operations in turn.

In a structured proof, it is permissible to make an arbitrary assumption in any subproof. The assumptions need not be members of the initial premise set. Note that such assumptions cannot be used directly outside of the subproof, only as conditions in derived implications, so they do not contaminate the superproof or any unrelated subproofs.

For example, in the proof we just saw, we used this assumption operation in the nested subproof even though p was not among the given premises.

An ordinary rule of inference applies to a particular subproof of a structured proof if and only if there is an instance of the rule in which all of the premises occur earlier in the subproof or in some superproof of the subproof. Importantly, it is not permissible to use sentences in subproofs of that subproof or in other subproofs of its superproofs.

For example, in the structured proof we have been looking at, it is okay to apply Implication Elimination to 1 and 3. And it is okay to use Implication Elimination on lines 2 and 4.

However, it is *not* acceptable to use a sentence from a subproof in applying an ordinary rule of inference in a superproof.

The last line of the malformed proof shown below gives an example of this. It is *not* permissible to use Implication Elimination as shown here because it uses a conclusion from a subproof as a premise in an application of an ordinary rule of inference in its superproof.

	1.	$p \Rightarrow q$	Premise
	2.	$q \Rightarrow r$	Premise
	3.	p	Assumption
	4.	q	Implication Elimination: 1, 3
	5.	r	Implication Elimination: 2, 4
	6.	$p \Rightarrow r$	Implication Introduction: 3, 5
Wrong!	7.	r	Implication Elimination: 2, 4 **Wrong!**

The malformed proof shown below is another example. Here, line 8 is illegal because line 4 is not in the current subproof or a superproof of this subproof.

$$
\begin{array}{rll}
1. & p \Rightarrow q & \text{Premise} \\
2. & q \Rightarrow r & \text{Premise} \\
3. & \quad p & \text{Assumption} \\
4. & \quad\quad q & \text{Implication Elimination: 1, 3} \\
5. & \quad\quad r & \text{Implication Elimination: 2, 4} \\
6. & p \Rightarrow r & \text{Implication Introduction: 3, 5} \\
7. & \quad\quad \neg r & \text{Assumption} \\
8. & \quad r & \text{Implication Elimination: 2, 4} \\
9. & \neg r \Rightarrow r & \text{Implication Introduction: 7, 8}
\end{array}
$$

Wrong! (left of line 8) **Wrong!** (right of line 8)

Correctly utilizing results derived in subproofs is the responsibility of a new type of rule of inference. Like an ordinary rule of inference, a structured rule of inference is a pattern of reasoning consisting of one or more premises and one or more conclusions. As before, the premises and conclusions can be schemas. However, the premises can also include conditions of the form $\varphi \mathbin{|-} \psi$, as in the following example. The rule in this case is called Implication Introduction, because it allows us to introduce new implications.

$$
\frac{\varphi \mathbin{|-} \psi}{\varphi \Rightarrow \psi}
$$

Once again, looking at the correct example above, we see that there is an instance of Implication Introduction (shown here on the left) in deriving line 6 from the subproof on lines 3–5. The application of Implication Introduction in the malformed proof is also okay in deriving line 7 from the subproof in lines 4–6.

Finally, we define a *structured proof* of a conclusion from a set of premises to be a sequence of (possibly nested) sentences terminating in an occurrence of the conclusion at the *top level* of the proof. Each step in the proof must be either (1) a premise (at the top level) or an assumption (other than at the top level) or (2) the result of applying an ordinary or structured rule of inference to earlier items in the sequence (subject to the constraints given above).

4.4   FITCH

Fitch is a proof system that is particularly popular in the Logic community. It is as powerful as many other proof systems and is far simpler to use. Fitch achieves this simplicity through its support for structured proofs and its use of structured rules of inference in addition to ordinary rules of inference.

Fitch has ten rules of inference in all. Nine of these are ordinary rules of inference. The other rule (Implication Introduction) is a structured rule of inference.

And Introduction (shown below on the left) allows us to derive a conjunction from its conjuncts. If a proof contains sentences φ_1 through φ_n, then we can infer their conjunction. *And*

Elimination (shown below on the right) allows us to derive conjuncts from a conjunction. If we have the conjunction of φ_1 through φ_n, then we can infer any of the conjuncts.

And Introduction

φ_1

...

φ_n
_____
$\varphi_1 \wedge ... \wedge \varphi_n$

And Elimination

$\varphi_1 \wedge ... \wedge \varphi_n$
_____
φ_i

Or Introduction allows us to infer an arbitrary disjunction so long as at least one of the disjuncts is already in the proof. *Or Elimination* is a little more complicated than And Elimination. Since we do not know which of the disjuncts is true, we cannot just drop the $\vee$. However, if we know that every disjunct entails some sentence, then we can infer that sentence even if we do not know which disjunct is true.

Or Introduction

φ_i
_____
$\varphi_1 \vee ... \vee \varphi_n$

Or Elimination

$\varphi_1 \vee ... \vee \varphi_n$

$\varphi_1 \Rightarrow \psi$

...

$\varphi_n \Rightarrow \psi$
_____
ψ

Negation Introduction allows us to derive the negation of a sentence if it leads to a contradiction. If we believe $(\varphi \Rightarrow \psi)$ and $(\varphi \Rightarrow \neg\psi)$, then we can derive that φ is false. *Negation Elimination* allows us to delete double negatives.

Negation Introduction

$\varphi \Rightarrow \psi$

$\varphi \Rightarrow \neg\psi$
_____
$\neg\varphi$

Negation Elimination

$\neg\neg\varphi$
_____
φ

Implication Introduction is the structured rule we saw in Section 4.3. If, by assuming φ, we can derive ψ, then we can derive $(\varphi \Rightarrow \psi)$. *Implication Elimination* is the first rule we saw Section 4.2.

Implication Introduction

$\varphi \vdash \psi$
_____
$\varphi \Rightarrow \psi$

Implication Elimination

$\varphi \Rightarrow \psi$

φ
_____
ψ

Biconditional Introduction allows us to deduce a biconditional from an implication and its inverse. *Biconditional Elimination* goes the other way, allowing us to deduce two implications from a single biconditional.

Biconditional Introduction

$$\frac{\begin{array}{l} \varphi \Rightarrow \psi \\ \psi \Rightarrow \varphi \end{array}}{\varphi \Leftrightarrow \psi}$$

Biconditional Elimination

$$\frac{\varphi \Leftrightarrow \psi}{\begin{array}{l} \varphi \Rightarrow \psi \\ \psi \Rightarrow \varphi \end{array}}$$

In addition to these rules of inference, it is common to include in Fitch proof editors several additional operations that are of use in constructing Fitch proofs. For example, the Premise operation allows one to add a new premise to a proof. The Reiteration operation allows one to reproduce an earlier conclusion for the purposes of clarity. Finally, the Delete operation allows one to delete unnecessary lines.

4.5   REASONING TIPS

The Fitch rules are all fairly simple to use; and, as we discuss in the next section, they are all that we need to prove any result that follows logically from any set of premises. Unfortunately, figuring out which rules to use in any given situation is not always that simple. Fortunately, there are a few tricks that help in many cases.

If the goal has the form $(\varphi \Rightarrow \psi)$, it is often good to assume φ and prove ψ and then use Implication Introduction to derive the goal. For example, if we have a premise q and we want to prove $(p \Rightarrow q)$, we assume p, reiterate q, and then use Implication Introduction to derive the goal.

$$
\begin{array}{lll}
1. & q & \text{Premise} \\
2. & \quad p & \text{Assumption} \\
3. & \quad q & \text{Reiteration: 1} \\
4. & p \Rightarrow q & \text{Implication Introduction: 2, 3}
\end{array}
$$

If the goal has the form $(\varphi \wedge \psi)$, we first prove φ and then prove ψ and then use And Introduction to derive $(\varphi \wedge \psi)$.

If the goal has the form $(\varphi \vee \psi)$, all we need to do is to prove φ or prove ψ, but we do not need to prove both. Once we have proved either one, we can disjoin that with anything else whatsoever.

If the goal has the form $(\neg\varphi)$, it is often useful to assume φ and prove a contradiction, meaning that φ must be false. To do this, we assume φ and derive some sentence ψ leading to $(\varphi \Rightarrow \psi)$. We assume φ again and derive some sentence $\neg\psi$ leading to $(\varphi \Rightarrow \neg\psi)$. Finally, we use Negation Introduction to derive $\neg\varphi$ as desired.

More generally, whenever we want to prove a sentence φ of any sort, we can sometimes succeed by assuming $\neg\varphi$, proving a contradiction as just discussed and thereby deriving $\neg\neg\varphi$. We can then apply Negation Elimination to get φ.

The following two tips suggest useful things we can try based on the form of the premises and the goal or subgoal we are trying to prove.

If there is a premise of the form $(\varphi \Rightarrow \psi)$ and our goal is to prove ψ, then it is often useful to try proving φ. If we succeed, we can then use Implication Elimination to derive ψ.

If we have a premise $(\varphi \vee \psi)$ and our goal is to prove χ, then we should try proving $(\varphi \Rightarrow \chi)$ and $(\psi \Rightarrow \chi)$. If we succeed, we can then use Or Elimination to derive χ.

As an example of using these tips in constructing the proof, consider the following problem. We are given $p \vee q$ and $\neg p$, and we are asked to prove q. Since the goal is not an implication or a conjunction or a disjunction or a negation, only the last of the goal-based tips applies. Unfortunately, this does not help us in this case. Luckily, the second of the premise-based tips is relevant because we have a disjunction as a premise. To use this all we need is to prove $p \Rightarrow q$ and $q \Rightarrow q$. To prove $p \Rightarrow q$, we use the first goal-based tip. We assume p and try to prove q. To do this we use that last goal-based tip. We assume $\sim q$ and prove p. Then we assume $\sim q$ and prove $\neg p$. Since we have proved p and $\neg p$ from $\neg q$, we can infer q. Using Implication Introduction, we then have $p \Rightarrow q$. Proving $q \Rightarrow q$ is easy. Finally, we can apply or elimination to get the desired result.

1.	$p \mid q$	Premise
2.	$\neg p$	Premise
3.	$\quad p$	Assumption
4.	$\quad\quad \neg q$	Assumption
5.	$\quad\quad p$	Reiteration: 3
6.	$\quad \neg q \Rightarrow p$	Implication Introduction: 4, 5
7.	$\quad\quad \neg q$	Assumption
8.	$\quad\quad \neg p$	Reiteration: 2
9.	$\quad \neg q \Rightarrow \neg p$	Implication Introduction: 7, 8
10.	$\quad \neg\neg q$	Negation Introduction: 6, 9
11.	$\quad q$	Negation Elimination: 10
12.	$p \Rightarrow q$	Implication Introduction: 3, 11
13.	$\quad q$	Assumption
14.	$q \Rightarrow q$	Implication Introduction: 13
15.	q	Or Elimination: 1, 12, 14

In general, when trying to generate a proof, it is useful to apply the premise tips to derive conclusions. However, this often works only for very short proofs. For more complex proofs, it is often useful to think backward from the desired conclusion before starting to prove things from the premises in order to devise a strategy for approaching the proof. This often suggests subproblems to be solved. We can then work on these simpler subproblems and put the solutions together to produce a proofs for our overall conclusion.

4.6   SOUNDNESS AND COMPLETENESS

In talking about Logic, we now have two notions—logical entailment and provability. A set of premises logically entails a conclusion if and only if every truth assignment that satisfies the premises also satisfies the conclusion. A sentence is provable from a set of premises if and only if there is a finite proof of the conclusion from the premises.

The concepts are quite different. One is based on truth assignments; the other is based on symbolic manipulation of expressions. Yet, for the proof systems we have been examining, they are closely related.

We say that a proof system is *sound* if and only if every provable conclusion is logically entailed. In other words, if $\Delta \mid- \varphi$, then $\Delta \mid= \varphi$. We say that a proof system is *complete* if and only if every logical conclusion is provable. In other words, if $\Delta \mid= \varphi$, then $\Delta \mid- \varphi$.

The Fitch system is sound and complete for the full language. In other words, for this system, logical entailment and provability are identical. An arbitrary set of sentences Δ logically entails an arbitrary sentence φ if and only if φ is provable from Δ using Fitch.

The upshot of this result is significant. On large problems, the proof method often takes fewer steps than the truth table method. (Disclaimer: In the worst case, the proof method may take just as many or more steps to find an answer as the truth table method.) Moreover, proofs are usually much smaller than the corresponding truth tables. So writing an argument to convince others does not take as much space.

RECAP

A *pattern* is an expression satisfying the grammatical rules of our language except for the occurrence of *metavariables* in place of various subparts of the expression. An *instance* of a pattern is the expression obtained by substituting expressions of the appropriate sort for the metavariables in the pattern so that the result is a legal expression. A *rule of inference* is a pattern of reasoning consisting of one set of patterns, called *premises*, and a second set of schemas, called *conclusions*. A *linear proof* of a conclusion from a set of premises is a sequence of sentences terminating in the conclusion in which each item is either (1) a premise or (2) the result of applying a rule of inference to earlier items in sequence. If there exists a proof of a sentence φ from a set Δ of premises and the axiom schemas and rules of inference of a proof system, then φ is said to be *provable* from Δ (written as $\Delta \mid- \varphi$) and is called a *theorem* of Δ. *Fitch* is a powerful yet simple proof system that supports structured proofs. A proof system is *sound* if and only if every provable conclusion is logically entailed. A proof system is *complete* if and only if every logical conclusion is provable. Fitch is sound and complete for Propositional Logic.

4.7   EXERCISES

4.1.   Given p and q and $(p \wedge q \Rightarrow r)$, use the Fitch system to prove r.

4.2.   Given $(p \wedge q)$, use the Fitch system to prove $(q \vee r)$.

4.3. Given $p \Rightarrow q$ and $q \Leftrightarrow r$, use the Fitch system to prove $p \Rightarrow r$.

4.4. Given $p \Rightarrow q$ and $m \Rightarrow p \vee q$, use the Fitch System to prove $m \Rightarrow q$.

4.5. Given $p \Rightarrow (q \Rightarrow r)$, use the Fitch System to prove $(p \Rightarrow q) \Rightarrow (p \Rightarrow r)$.

4.6. Use the Fitch System to prove $p \Rightarrow (q \Rightarrow p)$.

4.7. Use the Fitch System to prove $(p \Rightarrow (q \Rightarrow r)) \Rightarrow ((p \Rightarrow q) \Rightarrow (p \Rightarrow r))$.

4.8. Use the Fitch System to prove $(\neg p \Rightarrow q) \Rightarrow ((\neg p \Rightarrow \neg q) \Rightarrow p)$.

4.9. Given p, use the Fitch System to prove $\neg\neg p$.

4.10. Given $p \Rightarrow q$, use the Fitch System to prove $\neg q \Rightarrow \neg p$.

4.11. Given $p \Rightarrow q$, use the Fitch System to prove $\neg p \vee q$.

4.12. Use the Fitch System to prove $((p \Rightarrow q) \Rightarrow p) \Rightarrow p$.

4.13. Given $\neg(p \vee q)$, use the Fitch system to prove $(\neg p \wedge \neg q)$.

4.14. Use the Fitch system to prove the tautology $(p \vee \neg p)$.

CHAPTER 5

Propositional Resolution

5.1 INTRODUCTION

Propositional Resolution is a powerful rule of inference for Propositional Logic. Using Propositional Resolution (without axiom schemata or other rules of inference), it is possible to build a theorem prover that is sound and complete for all of Propositional Logic. What's more, the search space using Propositional Resolution is much smaller than for standard Propositional Logic.

This chapter is devoted entirely to Propositional Resolution. We start with a look at clausal form, a variation of the language of Propositional Logic. We then examine the resolution rule itself. We close with some examples.

5.2 CLAUSAL FORM

Propositional Resolution works only on expressions in *clausal form*. Before the rule can be applied, the premises and conclusions must be converted to this form. Fortunately, as we shall see, there is a simple procedure for making this conversion.

A *literal* is either an atomic sentence or a negation of an atomic sentence. For example, if p is a logical constant, the following sentences are both literals.

$$p$$
$$\neg p$$

A *clausal sentence* is either a literal or a disjunction of literals. If p and q are logical constants, then the following are clausal sentences.

$$p$$
$$\neg p$$
$$\neg p \vee q$$

A *clause* is the set of literals in a clausal sentence. For example, the following sets are the clauses corresponding to the clausal sentences above.

$$\{p\}$$
$$\{\neg p\}$$
$$\{\neg p, q\}$$

Note that the empty set {} is also a clause. It is equivalent to an empty disjunction and, therefore, is unsatisfiable. As we shall see, it is a particularly important special case.

As mentioned earlier, there is a simple procedure for converting an arbitrary set of Propositional Logic sentences to an equivalent set of clauses. The conversion rules are summarized below and should be applied in order.

1. Implications (I):

$$\varphi \Rightarrow \psi \quad \rightarrow \quad \neg\varphi \vee \psi$$
$$\varphi \Leftarrow \psi \quad \rightarrow \quad \varphi \vee \neg\psi$$
$$\varphi \Leftrightarrow \psi \quad \rightarrow \quad (\neg\varphi \vee \psi) \wedge (\varphi \vee \neg\psi)$$

2. Negations (N):

$$\neg\neg\varphi \quad \rightarrow \quad \varphi$$
$$\neg(\varphi \wedge \psi) \quad \rightarrow \quad \neg\varphi \vee \neg\psi$$
$$\neg(\varphi \vee \psi) \quad \rightarrow \quad \neg\varphi \wedge \neg\psi$$

3. Distribution (D):

$$\varphi \vee (\psi \wedge \chi) \quad \rightarrow \quad (\varphi \vee \psi) \wedge (\varphi \vee \chi)$$
$$(\varphi \wedge \psi) \vee \chi \quad \rightarrow \quad (\varphi \vee \chi) \wedge (\psi \vee \chi)$$
$$\varphi \vee (\varphi_1 \vee ... \vee \varphi_n) \quad \rightarrow \quad \varphi \vee \varphi_1 \vee ... \vee \varphi_n$$
$$(\varphi_1 \vee ... \vee \varphi_n) \vee \varphi \quad \rightarrow \quad \varphi_1 \vee ... \vee \varphi_n \vee \varphi$$
$$\varphi \wedge (\varphi_1 \wedge ... \wedge \varphi_n) \quad \rightarrow \quad \varphi \wedge \varphi_1 \wedge ... \wedge \varphi_n$$
$$(\varphi_1 \wedge ... \wedge \varphi_n) \wedge \varphi \quad \rightarrow \quad \varphi_1 \wedge ... \wedge \varphi_n \wedge \varphi$$

4. Operators (O):

$$\varphi_1 \vee ... \vee \varphi \quad \rightarrow \quad \{\varphi_1, ..., \varphi_n\}$$
$$\varphi_1 \wedge ... \wedge \varphi_n \quad \rightarrow \quad \{\varphi_1\}, ..., \{\varphi_n\}$$

As an example, consider the job of converting the sentence $(g \wedge (r \Rightarrow f))$ to clausal form. The conversion process is shown below.

$$g \wedge (r \Rightarrow f)$$
$$\text{I} \quad g \wedge (\neg r \vee f)$$
$$\text{N} \quad g \wedge (\neg r \vee f)$$
$$\text{D} \quad g \wedge (\neg r \vee f)$$
$$\text{O} \quad \{g\}$$
$$\{\neg r, f\}$$

As a slightly more complicated case, consider the following conversion. We start with the same sentence except that, in this case, it is negated.

$$\neg(g \wedge (r \Rightarrow f))$$

I $\quad \neg(g \wedge (\neg r \vee f))$

N $\quad \neg g \vee \neg(\neg r \vee f)$

$$\neg g \vee (\neg\neg r \wedge \neg f)$$

$$\neg g \vee (r \wedge \neg f)$$

D $\quad (\neg g \vee r) \wedge (\neg g \vee \neg f)$

O $\quad \{\neg g, r\}$

$\quad\quad \{\neg g, \neg f\}$

Note that, even though the sentences in these two examples are similar to start with (disagreeing on just one $\neg$ operator), the results are quite different.

5.3 RESOLUTION PRINCIPLE

The idea of Propositional Resolution is simple. Suppose we have the clause $\{p, q\}$. In other words, we know that p is true or q is true. Suppose we also have the clause $\{\neg q, r\}$. In other words, we know that q is false or r is true. One clause contains q, and the other contains $\neg q$. If q is false, then by the first clause p must be true. If q is true, then, by the second clause, r must be true. Since q must be either true or false, then it must be the case that either p is true or r is true. So we should be able to derive the clause $\{p, r\}$.

This intuition is the basis for the rule of inference shown below. Given a clause containing a literal χ and another clause containing the literal $\neg\chi$, we can infer the clause consisting of all the literals of both clauses without the complementary pair. This rule of inference is called *Propositional Resolution* or the *Resolution Principle*.

$$\frac{\{\varphi_1, \dots, \chi, \dots, \varphi_m\}}{\{\psi_1, \dots, \neg\chi, \dots, \psi_n\}}{\{\varphi_1, \dots, \varphi_m, \psi_1, \dots, \psi_n\}}$$

The case we just discussed is an example. If we have the clause $\{p, q\}$ and we also have the clause $\{\neg q, r\}$, then we can derive the clause $\{p, r\}$ in a single step.

$$\frac{\{p, q\}}{\{\neg q, r\}}{\{p, r\}}$$

Note that, since clauses are sets, there cannot be two occurrences of any literal in a clause. Therefore, in drawing a conclusion from two clauses that share a literal, we merge the two occurrences into one, as in the following example.

$$\frac{\{\neg p, q\}}{\{p, q\}}{\{q\}}$$

If either of the clauses is a singleton set, we see that the number of literals in the result is less than the number of literals in the other clause. For example, from the clause $\{p, q, r\}$ and the singleton clause $\{\neg p\}$, we can derive the shorter clause $\{q, r\}$.

$$\frac{\begin{array}{c} \{p, q, r\} \\ \{\neg p\} \end{array}}{\{q, r\}}$$

Resolving two singleton clauses leads to the *empty clause*; i.e., the clause consisting of no literals at all, as shown below. The derivation of the empty clause means that the database contains a contradiction.

$$\frac{\begin{array}{c} \{p\} \\ \{\neg p\} \end{array}}{\{\,\}}$$

If two clauses resolve, they may have more than one resolvent because there can be more than one way in which to choose the resolvents. Consider the following deductions.

$$\frac{\begin{array}{c} \{p, q\} \\ \{\neg p, \neg q\} \end{array}}{\begin{array}{c} \{p, \neg p\} \\ \{q, \neg q\} \end{array}}$$

Note, however, when two clauses have multiple pairs of complementary literals, only *one pair* of literals may be resolved at a time. For example, the following is *not* a legal application of Propositional Resolution.

$$\frac{\begin{array}{c} \{p, q\} \\ \{\neg p, \neg q\} \ \text{Wrong!} \end{array}}{\{\,\}}$$

If we were to allow this to go through, we would be saying these two clauses are inconsistent. However, it is perfectly possible for $(p \vee q)$ to be true and $(\neg p \vee \neg q)$ to be true at the same time. For example, we just let p be true and q be false, and we have satisfied both clauses.

It is noteworthy that resolution subsumes many of our other rules of inference. Consider, for example, Implication Elimination, shown below on the left. If we have $(p \Rightarrow q)$ and we have p, then we can deduce q. The clausal form of the premises and conclusion are shown below on the right. The implication $(p \Rightarrow q)$ corresponds to the clause $\{\neg p, q\}$, and p corresponds to the singleton clause $\{p\}$. We have two clauses with a complementary literal, and so we cancel the complementary literals and derive the clause $\{q\}$, which is the clausal form of q.

$$\frac{\begin{array}{c} p \Rightarrow q \\ p \end{array}}{q} \qquad\qquad \frac{\begin{array}{c} \{\neg p, q\} \\ \{p\} \end{array}}{\{q\}}$$

As another example, recall the example of formal reasoning introduced in Chapter 1. We said that, whenever we have two rules in which the left hand side of one contains a proposition constant that occurs on the right hand side of the other, then we can cancel those constants and deduce a new rule by combining the remaining constants on the left hand sides of both rules and the remaining constants on the right hand sides of both rules. As it turns out, this is just Propositional Resolution.

Recall that we illustrated this rule with the deduction shown below on the left. Given ($m \Rightarrow p \vee q$) and ($p \Rightarrow q$), we deduce ($m \Rightarrow q$). On the right, we have the clausal form of the sentences on the left. In place of the first sentence, we have the clause $\{\neg m, p, q\}$; and, in place of the second sentence, we have $\{\neg p, q\}$. Using resolution, we can deduce $\{\neg m, q\}$, which is the clausal form of the sentence we derived on the left.

$$
\begin{array}{ll}
m \Rightarrow p \vee q & \{\neg m, p, q\} \\
\underline{p \Rightarrow q} & \underline{\{\neg p, q\}} \\
m \Rightarrow q & \{\neg m, q\}
\end{array}
$$

5.4   RESOLUTION REASONING

Reasoning with the Resolution Principle is analogous to reasoning with other rules of inference. We start with premises; we apply the Resolution Principle to those premises; we apply the rule to the results of those applications; and so forth until we get to our desired conclusion or we run out of things to do.

Formally, we define a *resolution derivation* of a conclusion from a set of premises to be a finite sequence of clauses terminating in the conclusion in which each clause is either a premise or the result of applying the Resolution Principle to earlier members of the sequence.

Note that our definition of resolution derivation is analogous to our definition of linear proof. However, in this case, we do not use the word *proof*, because we reserve that word for a slightly different concept, which is discussed below.

In many cases, it is possible to find resolution derivations of conclusions from premises. Suppose, for example, we are given the clauses $\{\neg p, r\}$ and $\{\neg q, r\}$ and $\{p, q\}$. Then we can derive the conclusion $\{r\}$ as shown below.

$$
\begin{array}{lll}
1. & \{\neg p, r\} & \text{Premise} \\
2. & \{\neg q, r\} & \text{Premise} \\
3. & \{p, q\} & \text{Premise} \\
4. & \{q, r\} & 1, 3 \\
5. & \{r\} & 2, 4
\end{array}
$$

It is noteworthy that the resolution is not *generatively complete*, i.e., it is not possible to find resolution derivations for all clauses that are logically entailed by a set of premise clauses.

For example, given the clause $\{p\}$ and the clause $\{q\}$, there is no resolution derivation of $\{p, q\}$, despite the fact that it is logically entailed by the premises in this case.

As another example, consider that valid clauses (such as $\{p, \neg p\}$) are always true, and so they are logically entailed by any set of premises, including the empty set. However, Propositional Resolution requires some premises to have any effect. Given an empty set of premises, we would not be able to derive any conclusions, including these valid clauses.

On the other hand, we can be sure of one thing. If a set Δ of clauses is unsatisfiable, then there is guaranteed to be a resolution derivation of the empty clause from Δ. More generally, if a set Δ of Propositional Logic sentences is unsatisfiable, then there is guaranteed to be a resolution derivation of the empty clause from the clausal form of Δ.

As an example, consider the clauses $\{p, q\}$, $\{p, \neg q\}$, $\{\neg p, q\}$, and $\{\neg p, \neg q\}$. There is no truth assignment that satisfies all four of these clauses. Consequently, starting with these clauses, we should be able to derive the empty clause; and we can. A resolution derivation is shown below.

1.	$\{p, q\}$	Premise
2.	$\{p, \neg q\}$	Premise
3.	$\{\neg p, q\}$	Premise
4.	$\{\neg p, \neg q\}$	Premise
5.	$\{p\}$	1, 2
6.	$\{\neg p\}$	3, 4
7.	$\{\}$	5, 6

The good news is that we can use the relationship between unsatisfiability and logical entailment to produce a method for determining logical entailment as well. Recall that the Unsatisfiability Theorem introduced in Chapter 3 tells that a set Δ of sentences logically entails a sentence φ if and only if the set of sentences $\Delta \cup \{\neg\varphi\}$ is inconsistent. So, to determine logical entailment, all we need to do is to negate our goal, add it to our premises, and use Resolution to determine whether the resulting set is unsatisfiable.

Let's capture this idea with some definitions. A *resolution proof* of a sentence φ from a set Δ of sentences is a resolution derivation of the empty clause from the clausal form of $\Delta \cup \{\neg\varphi\}$. A sentence φ is *provable* from a set of sentences Δ by Propositional Resolution (written $\Delta \vdash \varphi$) if and only if there is a resolution proof of φ from Δ.

As an example of a resolution proof, consider one of the problems we saw earlier. We have three premises—p, $(p \Rightarrow q)$, and $(p \Rightarrow q) \Rightarrow (q \Rightarrow r)$. Our job is to prove r. A resolution proof is shown below. The first two clauses in the proof correspond to the first two premises of the problem. The third and fourth clauses in the proof correspond to the third premise. The fifth clause comes from the negation of the goal. Resolving the first clause with the second, we get the clause q, shown on line 6. Resolving this with the fourth clause gives us r. And resolving this with the clause on line 5 gives us the empty clause.

$$
\begin{array}{lll}
1. & \{p\} & \text{Premise} \\
2. & \{\neg p, q\} & \text{Premise} \\
3. & \{p, \neg q, r\} & \text{Premise} \\
4. & \{\neg q, r\} & \text{Premise} \\
5. & \{\neg r\} & \text{Premise} \\
6. & \{q\} & 1, 2 \\
7. & \{r\} & 4, 6 \\
8. & \{\} & 5, 7 \\
\end{array}
$$

Here is another example, this time illustrating the way in which we can use Resolution to prove valid sentences. Let's say that we have no premises at all and we want to prove $(p \Rightarrow (q \Rightarrow p))$, an instance of the Implication Creation axiom schema.

The first step is to negate this sentence and convert to clausal form. A trace of the conversion process is shown below. Note that we end up with three clauses.

$$
\begin{array}{ll}
 & \neg(p \Rightarrow (q \Rightarrow p)) \\
\text{I} & \neg(\neg p \vee (\neg q \vee p)) \\
\text{N} & \neg\neg p \wedge \neg(\neg q \vee p) \\
 & p \wedge (\neg\neg q \wedge \neg p) \\
\text{D} & p \wedge q \wedge \neg p \\
\text{O} & \{p\} \\
 & \{q\} \\
 & \{\neg p\} \\
\end{array}
$$

Finally, we take these clauses and produce a resolution derivation of the empty clause in one step.

$$
\begin{array}{lll}
1. & \{p\} & \text{Premise} \\
2. & \{q\} & \text{Premise} \\
3. & \{\neg p\} & \text{Premise} \\
4. & \{\} & 1, 3 \\
\end{array}
$$

One of the best features of Propositional Resolution is that it much more focussed than the other proof methods we have seen. There is no need to choose instantiations carefully or to search through infinitely many possible instantiations for rules of inference.

Moreover, unlike the other methods we have seen, Propositional Resolution can be used in a proof procedure that always terminates without losing completeness. Since there are only finitely many clauses that can be constructed from a finite set of proposition constants, the procedure eventually runs out of new conclusions to draw, and when this happens we can terminate our search for a proof.

RECAP

Propositional Resolution is a rule of inference for Propositional Logic. Propositional Resolution works only on expressions in *clausal form*. A *literal* is either an atomic sentence or a negation of an atomic sentence. A *clausal sentence* is either a literal or a disjunction of literals. A *clause* is the set of literals in a clausal sentence. The empty set {} is also a clause; it is equivalent to an empty disjunction and, therefore, is unsatisfiable. Given a clause containing a literal χ and another clause containing the literal $\neg\chi$, we can infer the clause consisting of all the literals of both clauses without the complementary pair. This rule of inference is called *Propositional Resolution* or the *Resolution Principle*. A *resolution derivation* of a conclusion from a set of premises is a finite sequence of clauses terminating in the conclusion in which each clause is either a premise or the result of applying the Resolution Principle to earlier members of the sequence. A *resolution proof* of a sentence φ from a set Δ of sentences is a resolution derivation of the empty clause from the clausal form of $\Delta \cup \{\neg\varphi\}$. A sentence φ is *provable* from a set of sentences Δ by Propositional Resolution (written $\Delta \vdash \varphi$) if and only if there is a resolution proof of φ from Δ. Resolution is not *generatively complete*, i.e., it is not possible to find resolution derivations for all clauses that are logically entailed by a set of premise clauses. On the other hand, it is *complete* in another sense— if a set Δ of clauses is unsatisfiable, then there is guaranteed to be a resolution derivation of the empty clause from Δ. More generally, if a set Δ of Propositional Logic sentences is unsatisfiable, then there is guaranteed to be a resolution derivation of the empty clause from the clausal form of Δ. Propositional Resolution can be used in a proof procedure that always terminates without losing completeness.

5.5   EXERCISES

5.1.   Convert the following sentences to clausal form.

(a) $p \wedge q \Rightarrow r \vee s$
(b) $p \vee q \Rightarrow r \vee s$
(c) $\neg(p \vee q \vee r)$
(d) $\neg(p \wedge q \wedge r)$
(e) $p \wedge q \Leftrightarrow r$

5.2.   What are the results of applying Propositional Resolution to the following pairs of clauses.

(a) $\{p, q, \neg r\}$ and $\{r, s\}$
(b) $\{p, q, r\}$ and $\{r, \neg s, \neg t\}$
(c) $\{q, \neg q\}$ and $\{q, \neg q\}$
(d) $\{\neg p, q, r\}$ and $\{p, \neg q, \neg r\}$

5.3.   Use Propositional Resolution to show that the clauses $\{p, q\}, \{\neg p, r\}, \{\neg p, \neg r\}, \{p, \neg q\}$ are not simultaneously satisfiable.

5.4. Given the premises $(p \Rightarrow q)$ and $(r \Rightarrow s)$, use Propositional Resolution to prove the conclusion $(p \vee r \Rightarrow q \vee s)$.

CHAPTER 6

Relational Logic

6.1 INTRODUCTION

Propositional Logic allows us to talk about relationships among individual propositions, and it gives us the machinery to derive logical conclusions based on these relationships. Suppose, for example, we believe that, if Jack knows Jill, then Jill knows Jack. Suppose we also believe that Jack knows Jill. From these two facts, we can conclude that Jill knows Jack using a simple application of Implication Elimination.

Unfortunately, when we want to say things more generally, we find that Propositional Logic is inadequate. Suppose, for example, that we wanted to say that, in general, if one person knows a second person, then the second person knows the first. Suppose, as before, that we believe that Jack knows Jill. How do we express the general fact in a way that allows us to conclude that Jill knows Jack? Here, Propositional Logic is inadequate; it gives us no way of succinctly encoding this more general belief in a form that captures its full meaning and allows us to derive such conclusions.

Relational Logic is an alternative to Propositional Logic that solves this problem. The trick is to augment our language with two new linguistic features, viz. *variables* and *quantifiers*. With these new features, we can express information about multiple objects without enumerating those objects; and we can express the existence of objects that satisfy specified conditions without saying which objects they are.

In this chapter, we proceed through the same stages as in the introduction to Propositional Logic. We start with syntax and semantics. We then discuss evaluation and satisfaction. We look at some examples. Then, we talk about properties of Relational Logic sentences and logical entailment for Relational Logic. Finally, we say a few words about the equivalence of Relational Logic and Propositional Logic and its decidability.

6.2 SYNTAX

In Propositional Logic, sentences are constructed from a basic vocabulary of propositional constants. In Relational Logic, there are no propositional constants; instead we have *object constants*, *relation constants*, and *variables*.

In our examples here, we write both variables and constants as strings of letters, digits, and a few non-alphanumeric characters (e.g., "_"). By convention, variables begin with letters from the end of the alphabet (viz. u, v, w, x, y, z). Examples include x, ya, and z_2. By convention, all constants begin with either alphabetic letters (other than u, v, w, x, y, z) or digits. Examples include a, b, 123, *comp225*, and *barack_obama*.

Note that there is no distinction in spelling between object constants and relation constants. The type of each such word is determined by its usage or, in some cases, in an explicit specification.

As we shall see, relation constants are used in forming complex expressions by combining them with an appropriate number of arguments. Accordingly, each relation constant has an associated *arity*, i.e., the number of arguments with which that relation constant can be combined. A relation constant that can combined with a single argument is said to be *unary*; one that can be combined with two arguments is said to be *binary*; one that can be combined with three arguments is said to be *ternary*; more generally, a relation constant that can be combined with n arguments is said to be n-ary.

A *vocabulary* consists of a set of object constants, a set of relation constants, and an assignment of arities for each of the relation constants in the vocabulary. (Note that this definition here is slightly non-traditional. In many textbooks, a vocabulary (sometimes called a *signature*) includes a specification of relation constants but not object constants, whereas our definition here includes both types of constants.)

A *term* is defined to be a variable or an object constant. Terms typically denote objects presumed or hypothesized to exist in the world; and, as such, they are analogous to noun phrases in natural language, e.g., *Joe* or *someone*.

There are three types of *sentences* in Relational Logic, viz. relational sentences (the analog of propositions in Propositional Logic), logical sentences (analogous to the logical sentences in Propositional Logic), and quantified sentences (which have no analog in Propositional Logic).

A *relational sentence* is an expression formed from an n-ary relation constant and n terms. For example, if q is a relation constant with arity 2 and if a and y are terms, then the expression shown below is a syntactically legal relational sentence. Relational sentences are sometimes called *atoms* to distinguish them from logical and quantified sentences.

$$q(a, y)$$

Logical sentences are defined as in Propositional Logic. There are negations, conjunctions, disjunctions, implications, and equivalences. See below for examples.

Negation:	$(\neg p(a))$
Conjunction:	$(p(a) \wedge q(b, c))$
Disjunction:	$(p(a) \vee q(b, c))$
Implication:	$(p(a) \Rightarrow q(b, c))$
Biconditional:	$(p(a) \Leftrightarrow q(b, c))$

Note that the syntax here is exactly the same as in Propositional Logic except that the elementary components are relational sentences rather than proposition constants.

Quantified sentences are formed from a *quantifier*, a variable, and an embedded sentence. The embedded sentence is called the *scope* of the quantifier. There are two types of quantified sentences in Relational Logic, viz. universally quantified sentences and existentially quantified sentences.

A *universally quantified sentence* is used to assert that all objects have a certain property. For example, the following expression is a universally quantified sentence asserting that, if p holds of an object, then q holds of that object and itself.

$$(\forall x.(p(x) \Rightarrow q(x,x)))$$

An *existentially quantified sentence* is used to assert that some object has a certain property. For example, the following expression is an existentially quantified sentence asserting that there is an object that satisfies p and, when paired with itself, satisfies q as well.

$$(\exists x.(p(x) \wedge q(x,x)))$$

Note that quantified sentences can be nested within other sentences. For example, in the first sentence below, we have quantified sentences inside of a disjunction. In the second sentence, we have a quantified sentence nested inside of another quantified sentence.

$$((\forall x.p(x)) \vee (\exists x.q(x,x)))$$
$$(\forall x.(\exists y.q(x,y)))$$

As with Propositional Logic, we can drop unneeded parentheses in Relational Logic, relying on precedence to disambiguate the structure of unparenthesized sentences. In Relational Logic, the precedence relations of the logical operators are the same as in Propositional Logic, and quantifiers have higher precedence than logical operators.

The following examples show how to parenthesize sentences with both quantifiers and logical operators. The sentences on the right are partially parenthesized versions of the sentences on the left. (To be fully parenthesized, we would need to add parentheses around each of the sentences as a whole.)

$$\forall x.p(x) \Rightarrow q(x) \qquad (\forall x.p(x)) \Rightarrow q(x)$$
$$\exists x.p(x) \wedge q(x) \qquad (\exists x.p(x)) \wedge q(x)$$

Notice that, in each of these examples, the quantifier does *not* apply to the second relational sentence, even though, in each case, that sentence contains an occurrence of the variable being quantified. If we want to apply the quantifier to a logical sentence, we must enclose that sentence in parentheses, as in the following examples.

$$\forall x.(p(x) \Rightarrow q(x))$$
$$\exists x.(p(x) \wedge q(x))$$

An expression in Relational Logic is *ground* if and only if it contains no variables. For example, the sentence $p(a)$ is ground, whereas the sentence $\forall x.p(x)$ is not.

An occurrence of a variable is *free* if and only if it is not in the scope of a quantifier of that variable. Otherwise, it is *bound*. For example, y is free and x is bound in the following sentence.

$$\exists x.q(x,y)$$

A sentence is *open* if and only if it has free variables. Otherwise, it is *closed*. For example, the first sentence below is open and the second is closed.

$$p(y) \Rightarrow \exists x.q(x,y)$$
$$\forall y.(p(y) \Rightarrow \exists x.q(x,y))$$

6.3   SEMANTICS

The semantics of Relational Logic presented here is termed *Herbrand semantics*. It is named after the logician Herbrand, who developed some of its key concepts. As Herbrand is French, it should properly be pronounced "air-brahn". However, most people resort to the Anglicization of this, instead pronouncing it "her-brand". (One exception is Stanley Peters, who has been known at times to pronounce it "hare-brained".)

The *Herbrand base* for a vocabulary is the set of all ground relational sentences that can be formed from the constants of the language. Said another way, it is the set of all sentences of the form $r(t_1,...,t_n)$, where r is an n-ary relation constant and $t_1, ... , t_n$ are object constants.

For a vocabulary with object constants a and b and relation constants p and q where p has arity 1 and q has arity 2, the Herbrand base is shown below.

$$\{p(a), p(b), q(a,a), q(a,b), q(b,a), q(b,b)\}$$

It is worthwhile to note that, for a given relation constant and a finite set of terms, there is an upper bound on the number of ground relational sentences that can be formed using that relation constant. In particular, for a set of terms of size b, there are b^n distinct n-tuples of object constants; and hence there are b^n ground relational sentences for each n-ary relation constant. Since the number of relation constants in a vocabulary is finite, this means that the Herbrand base is also finite.

A *truth assignment* for a Relational Logic language is a function that maps each ground relational sentence in the Herbrand base to a truth value. As in Propositional Logic, we use the digit 1 as a synonym for true and 0 as a synonym for false; and we refer to the value assigned to a ground relational sentence by writing the relational sentence with the name of the truth assignment as a superscript. For example, the truth assignment shown below is an example for the case of the language mentioned a few paragraphs above.

$$
\begin{array}{ccc}
p(a) & \rightarrow & 1 \\
p(b) & \rightarrow & 0 \\
q(a,a) & \rightarrow & 1 \\
q(a,b) & \rightarrow & 0 \\
q(b,a) & \rightarrow & 1 \\
q(b,b) & \rightarrow & 0
\end{array}
$$

As with Propositional Logic, once we have a truth assignment for the ground relational sentences of a language, the semantics of our operators prescribes a unique extension of that assignment to the complex sentences of the language.

The rules for logical sentences in Relational Logic are the same as those for logical sentences in Propositional Logic. A truth assignment satisfies a negation $\neg\varphi$ if and only if it does not satisfy φ. A truth assignment satisfies a conjunction $(\varphi_1 \wedge ... \wedge \varphi_n)$ if and only if it satisfies every φ_i. A truth assignment satisfies a disjunction $(\varphi_1 \vee ... \vee \varphi_n)$ if and only if it satisfies at least one φ_i. A truth assignment satisfies an implication $(\varphi \Rightarrow \psi)$ if and only if it does not satisfy φ or does satisfy ψ. A truth assignment satisfies an equivalence $(\varphi \Leftrightarrow \psi)$ if and only if it satisfies both φ and ψ or it satisfies neither φ nor ψ.

In order to define satisfaction of quantified sentences, we need the notion of instances. An *instance* of an expression is an expression in which all free variables have been consistently replaced by ground terms. Consistent replacement here means that, if one occurrence of a variable is replaced by a ground term, then all occurrences of that variable are replaced by the same ground term.

A universally quantified sentence is true for a truth assignment if and only if *every* instance of the scope of the quantified sentence is true for that assignment. An existentially quantified sentence is true for a truth assignment if and only if *some* instance of the scope of the quantified sentence is true for that assignment.

A truth assignment *satisfies* a sentence with free variables if and only if it satisfies every instance of that sentence. A truth assignment *satisfies* a set of sentences if and only if it satisfies every sentence in the set.

6.4   EVALUATION

Evaluation for Relational Logic is similar to evaluation for Propositional Logic. The only difference is that we need to deal with quantifiers. In order to evaluate a universally quantified sentence, we check that all instances of the scope are true. (We are in effect treating it as the conjunction of all those instances.) In order to evaluate an existentially quantified sentence, we check that at least one instance of the scope. (We are in effect treating it as the disjunction of those instances.)

Once again, assume we have a language with a unary relation constant p, a binary relation constant q, and object constants a and b; and consider our truth assignment from the last section.

What is the truth value of the sentence $\forall x.(p(x) \Rightarrow q(x,x))$ under this assignment? There are two instances of the scope of this sentence. See below.

$$p(a) \Rightarrow q(a,a)$$
$$p(b) \Rightarrow q(b,b)$$

We know that $p(a)$ is true and $q(a,a)$ is true, so the first instance is true. $q(b,b)$ is false, but so is $p(b)$ so the second instance is true as well.

$$(p(a) \Rightarrow q(a,a)) \quad \rightarrow \quad 1$$
$$(p(b) \Rightarrow q(b,b)) \quad \rightarrow \quad 1$$

Since both instances are true, the original quantified sentence is true as well.

$$\forall x.(p(x) \Rightarrow q(x,x)) \rightarrow 1$$

Now let's consider a case with nested quantifiers. Is $\forall x.\exists y.q(x,y)$ true or false for the truth assignment shown above? As before, we know that this sentence is true if every instance of its scope is true. The two possible instances are shown below.

$$\exists y.q(a,y)$$
$$\exists y.q(b,y)$$

To determine the truth of $\exists y.q(a,y)$, we must find at least one instance of $q(a,y)$ that is true. The possibilities are shown below.

$$q(a,a)$$
$$q(a,b)$$

Looking at our truth assignment, we see that the first of these is true and the second is false.

$$q(a,a) \quad \rightarrow \quad 1$$
$$q(a,b) \quad \rightarrow \quad 0$$

Since one of these instances is true, the existential sentence as a whole is true.

$$\exists y.q(a,y) \rightarrow 1$$

Now, we do the same for the second existentially quantified. The possible instances in this case are shown below.

$$q(b,a)$$
$$q(b,b)$$

Of these, the first is true, and the second is false.

$$q(b,a) \quad \rightarrow \quad 1$$
$$q(b,b) \quad \rightarrow \quad 0$$

Again, since one of these instances is true, the existential sentence as a whole is true.

$$\exists y.q(b,y) \rightarrow 1$$

At this point, we have truth values for our two existential sentences. Since both instances of the scope of our original universal sentence are true, the sentence as a whole must be true as well.

$$\forall x.\exists y.q(x,y) \rightarrow 1$$

6.5   SATISFACTION

As in Propositional Logic, it is in principle possible to build a truth table for any set of sentences in Relational Logic. This truth table can then be used to determine which truth assignments satisfy a given set of sentences.

As an example, let us assume we have a language with just two object constants a and b and two unary relation constants p and q. Now consider the sentences shown below, and assume our job is to find a truth assignment that satisfies these sentences.

$$p(a) \lor p(b)$$
$$\forall x.(p(x) \Rightarrow q(x))$$
$$\exists x.q(x)$$

A truth table for this problem is shown below. Each of the four columns on the left represents one of the elements of the Herbrand base for this language. The three columns on the right represent our sentences.

$p(a)$	$p(b)$	$q(a)$	$q(b)$	$p(a) \lor p(b)$	$\forall x.(p(x) \Rightarrow q(x))$	$\exists x.q(x)$
1	1	1	1	1	1	1
1	1	1	0	1	0	1
1	1	0	1	1	0	1
1	1	0	0	1	0	0
1	0	1	1	1	1	1
1	0	1	0	1	1	1
1	0	0	1	1	0	1
1	0	0	0	1	0	0
0	1	1	1	1	1	1
0	1	1	0	1	0	1
0	1	0	1	1	1	1
0	1	0	0	1	0	0
0	0	1	1	0	1	1
0	0	1	0	0	1	1
0	0	0	1	0	1	1
0	0	0	0	0	1	0

Looking at the table, we see that there are twelve truth assignments that make the first sentence true, nine that make the second sentence true, twelve that make the third sentence true, and five that make them all true (rows 1, 5, 6, 9, and 11).

6.6 EXAMPLE–SORORITY WORLD

Consider once again the Sorority World example introduced in Chapter 1. Recall that this world focusses on the interpersonal relations of a small sorority. There are just four members—Abby, Bess, Cody, and Dana. Our goal is to represent information about who likes whom.

In order to encode this information in Relational Logic, we adopt a vocabulary with four object constants (*abby*, *bess*, *cody*, *dana*) and one binary relation constant (*likes*).

If we had complete information about the likes and dislikes of the girls, we could completely characterize the state of affairs as a set of ground relational sentences or negations of ground relational sentences, like the ones shown below, with one sentence for each member of the Herbrand base. (In our example here, we have written the positive literals in black and the negative literals in grey in order to distinguish the two more easily.)

¬*likes(abby,abby)*	¬*likes(abby,bess)*	*likes(abby,cody)*	¬*likes(abby,dana)*
¬*likes(bess,abby)*	¬*likes(bess,bess)*	*likes(bess,cody)*	¬*likes(bess,dana)*
likes(cody,abby)	*likes(cody,bess)*	¬*likes(cody,cody)*	*likes(cody,dana)*
¬*likes(dana,abby)*	¬*likes(dana,bess)*	*likes(dana,cody)*	¬*likes(dana,dana)*

To make things more interesting, let's assume that we do *not* have complete information, only fragments of information about the girls' likes and dislikes. Let's see how we can encode such fragments in Relational Logic.

Let's start with a simple disjunction. *Bess likes Cody or Dana.* Encoding a sentence with a disjunctive noun phrase (such as *Cody or Dana*) is facilitated by first rewriting the sentence as a disjunction of simple sentences. *Bess likes Cody or Bess likes Dana.* In Relational Logic, we can express this fact as a simple disjunction with the two possibilities as disjuncts.

$$likes(bess,cody) \lor likes(bess,dana)$$

Abby likes everyone Bess likes. Again, paraphrasing helps translate. *If Bess likes a girl, then Abby also likes her.* Since this is a fact about everyone, we use a universal quantifier.

$$\forall y.(likes(bess,y) \Rightarrow likes(abby,y))$$

Cody likes everyone who likes her. In other words, *if some girl likes Cody, then Cody likes that girl.* Again, we use a universal quantifier.

$$\forall y.(likes(y,cody) \Rightarrow likes(cody,y))$$

Bess likes somebody who likes her. The word *somebody* here is a tip-off that we need to use an existential quantifier.

$$\exists y.(likes(bess,y) \land likes(y,bess))$$

Nobody likes herself. The use of the word *nobody* here suggests a negation. A good technique in such cases is to rewrite the English sentence as the negation of a positive version of the sentence before translating to Relational Logic.

$$\neg\exists x.likes(x,x)$$

Everybody likes somebody. Here we have a case requiring two quantifiers, one universal and one existential. The key to this case is getting the quantifiers in the right order. Reversing them leads to a very different statement.

$$\forall x.\exists y.likes(x,y)$$

There is someone everyone likes. The preceding sentence tells us that everyone likes someone, but that someone can be different for different people. This sentence tells us that everybody likes the same person.

$$\exists y.\forall x.likes(x,y)$$

6.7   EXAMPLE–BLOCKS WORLD

The Blocks World is a popular application area for illustrating ideas in the field of Artificial Intelligence. A typical Blocks World scene is shown in Figure 6.1.

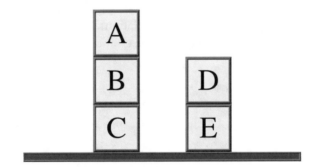

Figure 6.1: One state of Blocks World.

Most people looking at this figure interpret it as a configuration of five toy blocks. Some people conceptualize the table on which the blocks are resting as an object as well; but, for simplicity, we ignore it here.

In order to describe this scene, we adopt a vocabulary with five object constants, as shown below, with one object constant for each of the five blocks in the scene. The intent here is for each of these object constants to represent the block marked with the corresponding capital letter in the scene.

$$\{a, b, c, d, e\}$$

In a spatial conceptualization of the Blocks World, there are numerous meaningful relations. For example, it makes sense to think about the relation that holds between two blocks if and only if one is resting on the other. In what follows, we use the relation constant *on* to refer to this relation. We might also think about the relation that holds between two blocks if and only if one is anywhere above the other, i.e., the first is resting on the second or is resting on a block that is resting on the second, and so forth. In what follows, we use the relation constant *above* to talk about this relation. There is the relation that holds of three blocks that are stacked one on top of the other. We use the relation constant *stack* as a name for this relation. We use the relation constant *clear* to denote the relation that holds of a block if and only if there is no block on top of it. We use the relation constant *table* to denote the relation that holds of a block if and only if that block is resting on the table. The set of relation constants corresponding to this conceptualization is shown below.

$$\{on,\ above,\ stack,\ clear,\ table\}$$

The arities of these relation constants are determined by their intended use. Since *on* is intended to denote a relation between two blocks, it has arity 2. Similarly, *above* has arity 2. The *stack* relation constant has arity 3. Relation constants *clear* and *table* each have arity 1.

Given this vocabulary, we can describe the scene in Figure 6.1 by writing ground literals that state which relations hold of which objects or groups of objects. Let's start with *on*. The following sentences tell us directly for each ground relational sentence whether it is true or false. (Once again, we have written the positive literals in black and the negative literals in grey in order to distinguish the two more easily.)

$\neg on(a,a)$	$on(a,b)$	$\neg on(a,c)$	$\neg on(a,d)$	$\neg on(a,e)$
$\neg on(b,a)$	$\neg on(b,b)$	$on(b,c)$	$\neg on(b,d)$	$\neg on(b,e)$
$\neg on(c,a)$	$\neg on(c,b)$	$\neg on(c,c)$	$\neg on(c,d)$	$\neg on(c,e)$
$\neg on(d,a)$	$\neg on(d,b)$	$\neg on(d,c)$	$\neg on(d,d)$	$on(d,e)$
$\neg on(e,a)$	$\neg on(e,b)$	$\neg on(e,c)$	$\neg on(e,d)$	$\neg on(e,e)$

We can do the same for the other relations. However, there is an easier way. Each of the remaining relations can be defined in terms of *on*. These definitions together with our facts about the *on* relation logically entail every other ground relational sentence or its negation. Hence, given these definitions, we do not need to write out any additional data.

A block satisfies the *clear* relation if and only if there is nothing on it.

$$\forall y.(clear(y) \Leftrightarrow \neg \exists x.on(x,y))$$

A block satisfies the *table* relation if and only if it is not on some block.

$$\forall x.(table(x) \Leftrightarrow \neg \exists y.on(x,y))$$

Three blocks satisfy the *stack* relation if and only if the first is on the second and the second is on the third.

$$\forall x.\forall y.\forall z.(stack(x,y,z) \Leftrightarrow on(x,y) \wedge on(y,z))$$

The *above* relation is a bit tricky to define correctly. One block is above another block if and only if the first block is on the second block or it is on another block that is above the second block. Also, no block can be above itself. Given a complete definition for the *on* relation, these two axioms determine a unique *above* relation.

$$\forall x.\forall z.(above(x,z) \Leftrightarrow on(x,z) \vee \exists y.(on(x,y) \wedge above(y,z)))$$
$$\neg \exists x.above(x,x)$$

One advantage to defining relations in terms of other relations is economy. If we record *on* information for every object and encode the relationship between the *on* relation and these other relations, there is no need to record any ground relational sentences for those relations.

Another advantage is that these general sentences apply to Blocks World scenes other than the one pictured here. It is possible to create a Blocks World scene in which none of the *on* sentences we have listed is true, but these general definitions are still correct.

6.8   EXAMPLE–MODULAR ARITHMETIC

In this example, we show how to characterize Modular Arithmetic in Relational Logic. In Modular Arithmetic, there are only finitely many objects. For example, in Modular Arithmetic with modulus 4, we would have just four integers—0, 1, 2, 3—and that's all. Our goal here to define the addition relation. Admittedly, this is a modest goal; but, once we see how to do this; we can use the same approach to define other arithmetic relations.

Let's start with the *same* relation, which is true of every number and itself and is false for numbers that are different. We can completely characterize the *same* relation by writing ground relational sentences, one positive sentence for each number and itself and negative sentences for all of the other cases.

same(0,0)	¬*same(0,1)*	¬*same(0,2)*	¬*same(0,3)*
¬*same(1,0)*	*same(1,1)*	¬*same(1,2)*	¬*same(1,3)*
¬*same(2,0)*	¬*same(2,1)*	*same(2,2)*	¬*same(2,3)*
¬*same(3,0)*	¬*same(3,1)*	¬*same(3,2)*	*same(3,3)*

Now, let's axiomatize the *next* relation, which, for each number, gives the next larger number, wrapping back to 0 after we reach 3.

$$next(0,1)$$
$$next(1,2)$$
$$next(2,3)$$
$$next(3,0)$$

Properly, we should write out the negative literals as well. However, we can save that work by writing a single axiom asserting that *next* is a functional relation, i.e., for each member of the Herbrand base, there is just one successor.

$$\forall x.\forall y.\forall z.(next(x,y) \land next(x,z) \Rightarrow same(y,z))$$

In order to see why this saves us the work of writing out the negative literals, we can write this axiom in the equivalent form shown below.

$$\forall x.\forall y.\forall z.(next(x,y) \land \neg same(y,z) \Rightarrow \neg next(x,z))$$

The addition table for Modular Arithmetic is the usual addition table for arbitrary numbers except that we wrap around whenever we get past 3. For such a small arithmetic, it is easy to write out the ground relational sentences for addition, as shown below.

$plus(0,0,0)$	$plus(1,0,1)$	$plus(2,0,2)$	$plus(3,0,3)$
$plus(0,1,1)$	$plus(1,1,2)$	$plus(2,1,3)$	$plus(3,1,0)$
$plus(0,2,2)$	$plus(1,2,3)$	$plus(2,2,0)$	$plus(3,2,1)$
$plus(0,3,3)$	$plus(1,3,0)$	$plus(2,3,1)$	$plus(3,3,2)$

As with *next*, we avoid writing out the negative literals by writing a suitable functionality axiom for *plus*.

$$\forall x.\forall y.\forall z.\forall w.(plus(x,y,z) \land \neg same(z,w) \Rightarrow \neg plus(x,y,w))$$

That's one way to do things, but we can do better. Rather than writing out all of those relational sentences, we can use Relational Logic to define *plus* in terms of *same* and *next* and use that axiomatization to deduce the ground relational sentences. The definition is shown below. First, we have an identity axiom. Adding 0 to any number results in the same number. Second we have a successor axiom. If z is the sum of x and y, then the sum of the successor of x and y is the successor of z. Finally, we have our functionality axiom once again.

$\forall y.plus(0,y,y)$
$\forall x.\forall y.\forall z.\forall x2.\forall z2.(plus(x,y,z) \land next(x,x2) \land next(z,z2) \Rightarrow plus(x2,y,z2))$
$\forall x.\forall y.\forall z.\forall w.(plus(x,y,z) \land \neg same(z,w) \Rightarrow \neg plus(x,y,w))$

One advantage of doing things this way is economy. With these sentences, we do not need the ground relational sentences about *plus* given above. They are all logically entailed by our sentences about *next* and the definitional sentences. A second advantage is versatility. Our sentences define *plus* in terms of *next* for arithmetic with any modulus, not just modulus 4.

6.9   LOGICAL PROPERTIES

As we have seen, some sentences are true in some truth assignments and false in others. However, this is not always the case. There are sentences that are always true and sentences that are always false as well as sentences that are sometimes true and sometimes false.

As with Propositional Logic, this leads to a partition of sentences into three disjoint categories. A sentence is *valid* if and only if it is satisfied by *every* truth assignment. A sentence is *unsatisfiable* if and only if it is not satisfied by any truth assignment. A sentence is *contingent* if and only if there is some truth assignment that satisfies it and some truth assignment that falsifies it.

Alternatively, we can classify sentences into two overlapping categories. A sentence is *satisfiable* if and only if it is satisfied by at least one truth assignment, i.e., it is either valid or contingent. A sentence is *falsifiable* if and only if there is at least one truth assignment that makes it false, i.e., it is either contingent or unsatisfiable.

Note that these definitions are the same as in Propositional Logic. Moreover, some of our results are the same as well. If we think of ground relational sentences as propositions, we get similar results for the two logics - a ground sentence in Relational Logic is valid / contingent / unsatisfiable if and only if the corresponding sentence in Propositional Logic is valid / contingent / unsatisfiable.

Here, for example, are Relational Logic versions of common Propositional Logic validities—the Law of the Excluded Middle, Double Negation, and deMorgan's laws for distributing negation over conjunction and disjunction.

$$p(a) \vee \neg p(a)$$
$$p(a) \Leftrightarrow \neg\neg p(a)$$
$$\neg(p(a) \wedge q(a,b)) \Leftrightarrow (\neg p(a) \vee \neg q(a,b))$$
$$\neg(p(a) \vee q(a,b)) \Leftrightarrow (\neg p(a) \wedge \neg q(a,b))$$

Of course, not all sentences in Relational Logic are ground. There are valid sentences of Relational Logic for which there are no corresponding sentences in Propositional Logic.

The *Common Quantifier Reversal* tells us that reversing quantifiers of the same type has no effect on truth assignment.

$$\forall x.\forall y.q(x,y) \Leftrightarrow \forall y.\forall x.q(x,y)$$
$$\exists x.\exists y.q(x,y) \Leftrightarrow \exists y.\exists x.q(x,y))$$

Existential Distribution tells us that it is okay to move an existential quantifier inside of a universal quantifier. (Note that the reverse is not valid, as we shall see later.)

$$\exists y.\forall x.q(x,y) \Rightarrow \forall x.\exists y.q(x,y)$$

Finally, *Negation Distribution* tells us that it is okay to distribute negation over quantifiers of either type by flipping the quantifier and negating the scope of the quantified sentence.

$$\neg\forall x.p(x) \Leftrightarrow \exists x.\neg p(x)$$
$$\neg\exists x.p(x) \Leftrightarrow \forall x.\neg p(x)$$

6.10 LOGICAL ENTAILMENT

A set of Relational Logic sentences Δ *logically entails* a sentence φ (written $\Delta \models \varphi$) if and only if every truth assignment that satisfies Δ also satisfies φ.

As with validity and contingency and satisfiability, this definition is the same for Relational Logic as for Propositional Logic. As before, if we treat ground relational sentences as propositions, we get similar results. In particular, a set of ground premises in Relational Logic logically entails a ground conclusion in Relational Logic if and only if the corresponding set of Propositional Logic premises logically entails the corresponding Propositional Logic conclusion.

For sentences without variables, we have the following results. The sentence $p(a)$ logically entails $(p(a) \vee p(b))$. The sentence $p(a)$ does *not* logically entail $(p(a) \wedge p(b))$. However, any set of sentences containing both $p(a)$ and $p(b)$ does logically entail $(p(a) \wedge p(b))$.

The presence of variables allows for additional logical entailments. For example, the premise $\exists y.\forall x.q(x,y)$ logically entails the conclusion $\forall x.\exists y.q(x,y)$. If there is *some* object y that is paired with every x, then every x has some object that it pairs with, viz. y.

$$\exists y.\forall x.q(x,y) \models \forall x.\exists y.q(x,y)$$

Here is another example. The premise $\forall x.\forall y.q(x,y)$ logically entails the conclusion $\forall x.\forall y.q(y,x)$. The first sentence says that q is true for all pairs of objects, and the second sentence says the exact same thing. In cases like this, we can interchange variables.

$$\forall x.\forall y.q(x,y) \models \forall x.\forall y.q(y,x)$$

Understanding logical entailment for Relational Logic is complicated by the fact that it is possible to have free variables in Relational Logic sentences. Consider, for example, the premise $q(x,y)$ and the conclusion $q(y,x)$. Does $q(x,y)$ logically entail $q(y,x)$ or not?

Our definition for logical entailment and the semantics of Relational Logic give a clear answer to this question. Logical entailment holds if and only if every truth assignment that satisfies the premise satisfies the conclusion. A truth assignment satisfies a sentence with free variables if and only if it satisfies every instance. In other words, a sentence with free variables is equivalent to the sentence in which all of the free variables are universally quantified. In other words, $q(x,y)$ is satisfied if and only if $\forall x.\forall y.q(x,y)$ is satisfied, and similarly for $q(y,x)$. So, the first sentence here logically entails the second if and only if $\forall x.\forall y.q(x,y)$ logically entails $\forall x.\forall y.q(y,x)$; and, as we just saw, this is, in fact, the case.

6.11 RELATIONAL LOGIC AND PROPOSITIONAL LOGIC

One interesting feature of *Relational Logic* (RL) is that it is expressively equivalent to Propositional Logic (PL). For any RL language, we can produce a pairing between the ground relational sentences in that language and the proposition constants in a Propositional Logic language. Given this correspondence, for any set of *arbitrary* sentences in our RL language, there is a corresponding set of sentences in the language of PL such that any RL truth assignment that satisfies our RL sentences agrees with the corresponding Propositional Logic truth assignment applied to the Propositional Logic sentences.

The procedure for transforming our RL sentences to PL has multiple steps, but each step is easy. We first convert our sentences to prenex form, then we ground the result, and we rewrite in Propositional Logic. Let's look at these steps in turn.

A sentence is in *prenex form* if and only if it is closed and all of the quantifiers are on the outside of all logical operators.

Converting a set of RL sentences to a logically equivalent set in prenex form is simple. First, we rename variables in different quantified sentences to eliminate any duplicates. We then apply quantifier distribution rules in reverse to move quantifiers outside of logical operators. Finally, we universally quantify any free variables in our sentences.

For example, to convert the sentence $\forall y.p(x,y) \vee \exists y.q(y)$ to prenex form, we first rename one of the variables. In this case, let's rename the y in the second disjunct to z. This results in the sentence $\forall y.p(x,y) \vee \exists z.q(z)$. We then apply the distribution rules in reverse to produce $\forall y.\exists z.(p(x,y) \vee q(z))$. Finally, we universally quantify the free variable x to produce the prenex form of our original sentence, viz. $\forall x.\forall y.\exists z.(p(x,y) \vee q(z))$.

Once we have a set of sentences in prenex form, we compute the grounding. We start with our initial set Δ of sentences and we incrementally build up our grounding Γ. On each step we process a sentence in Δ, using the procedure described below. The procedure terminates when Δ becomes empty. The set Γ at that point is the grounding of the input.

(1) The first rule covers the case when the sentence φ being processed is ground. In this case, we remove the sentence from Delta and add it to Gamma.

$$\Delta_{i+1} = \Delta_i - \{\varphi\}$$
$$\Gamma_{i+1} = \Gamma_i \cup \{\varphi\}$$

(2) If our sentence is of the form $\forall v.\varphi[v]$, we eliminate the sentence from Δ_i and replace it with copies of the scope, one copy for each object constant τ in our language.

$$\Delta_{i+1} = \Delta_i - \{\forall v.\varphi[v]\} \cup \{\varphi[\tau] \mid \tau \text{ an object constant}\}$$
$$\Gamma_{i+1} = \Gamma_i$$

(3) If our sentence of the form $\exists v.\varphi[v]$, we eliminate the sentence from Δ_i and replace it with a disjunction, where each disjunct is a copy of the scope in which the quantified variable is replaced by an object constant in our language.

$$\Delta_{i+1} = \Delta_i - \{\exists v.\varphi[v]\} \cup \{\varphi[\tau_1] \vee ... \vee \varphi[\tau_n]\}$$
$$\Gamma_{i+1} = \Gamma_i$$

The procedure halts when Δ_i becomes empty. The set Γ_i is the grounding of the input. It is easy to see that Γ_i is logically equivalent to the input set.

Here is an example. Suppose we have a language with just two object constants a and b. And suppose we have the set of sentences shown below. We have one ground sentence, one universally quantified sentence, and one existentially quantified sentence. All are in prenex form.

$$\{p(a), \forall x.(p(x) \Rightarrow q(x)), \exists x.\neg q(x)\}$$

A trace of the procedure is shown below. The first sentence is ground, so we remove it from Δ add it to Γ. The second sentence is universally quantified, so we replace it with a copy for each of our two object constants. The resulting sentences are ground, and so we move them one by one from Δ to Γ. Finally, we ground the existential sentence and add the result to Δ and then move the ground sentence to Γ. At this point, since Δ is empty, Γ is our grounding.

$$\Delta_0 = \{\{p(a), \forall x.(p(x) \Rightarrow q(x)), \exists x.\neg q(x)\}$$
$$\Gamma_0 = \{\}$$

$$\Delta_1 = \{\forall x.(p(x) \Rightarrow q(x)), \exists x.\neg q(x)\}$$
$$\Gamma_1 = \{p(a)\}$$

$$\Delta_2 = \{p(a) \Rightarrow q(a), p(b) \Rightarrow q(b), \exists x.\neg q(x)\}$$
$$\Gamma_2 = \{p(a)\}$$

$$\Delta_3 = \{p(b) \Rightarrow q(b), \exists x.\neg q(x)\}$$
$$\Gamma_3 = \{p(a), p(a) \Rightarrow q(a)\}$$

$$\Delta_4 = \{\exists x.\neg q(x)\}$$
$$\Gamma_4 = \{p(a), p(a) \Rightarrow q(a), p(b) \Rightarrow q(b)\}$$

$$\Delta_5 = \{\neg q(a) \vee \neg q(b)\}$$
$$\Gamma_5 = \{p(a), p(a) \Rightarrow q(a), p(b) \Rightarrow q(b)\}$$

$$\Delta_6 = \{\}$$
$$\Gamma_6 = \{p(a), p(a) \Rightarrow q(a), p(b) \Rightarrow q(b), \neg q(a) \vee \neg q(b)\}$$

Once we have a grounding Γ, we replace each ground relational sentence in Γ by a proposition constant. The resulting sentences are all in Propositional Logic; and the set is equivalent to the sentences in Δ in that any RL truth assignment that satisfies our RL sentences agrees with the corresponding Propositional Logic truth assignment applied to the Propositional Logic sentences.

For example, let's represent the RL sentence $p(a)$ with the proposition pa; let's represent $p(b)$ with pb; let's represent $q(a)$ with qa; and let's represent $q(b)$ with qb. With this correspondence, we can represent the sentences in our grounding with the Propositional Logic sentences shown below.

$$\{pa, pa \Rightarrow qa, pb \Rightarrow qb, \neg qa \vee \neg qb\}$$

Since the question of unsatisfiability for PL is decidable, then the question of unsatisfiability for RL is also decidable. Since logical entailment and unsatisfiability are correlated, we also know that the question of logical entailment for RL is decidable.

Another consequence of this correspondence between RL and PL is that, like PL, RL is *compact*—every unsatisfiable set of sentences contains a finite subset that is unsatisfiable. This is important as it assures us that we can demonstrate the unsatisfiability by analyzing just a finite set of sentences; and, as we shall see in the next chapter, logical entailment can be demonstrated with finite proofs.

RECAP

Relational Logic is an alternative to Propositional Logic that includes some linguistic features, viz. constants and variables and quantifiers. In Relational Logic, simple sentences have more structure than in Propositional Logic. Furthermore, using variables and quantifiers, we can express information about multiple objects without enumerating those objects; and we can express the existence of objects that satisfy specified conditions without saying which objects they are. The syntax of Relational Logic begins with object constants and relation constants. *Relational sentences* are the atomic elements from which more complex sentences are built. *Logical sentences* are formed by combining simpler sentences with logical operators. In the version of Relational Logic used here, there are five types of logical sentences - negations, conjunctions, disjunctions, implications, and equivalences. There are two types of *quantified sentences*, viz. *universal sentences* and *existential sentences*. The *Herbrand base* for a Relational Logic language is the set of all ground relational sentences in the language. A *truth assignment* for a Relational Logic language is a mapping that assigns a truth value to each element of it Herbrand base. The truth or falsity of compound sentences is determined from a truth assignment using rules based on the five logical operators of the language. A truth assignment *satisfies* a sentence if and only if the sentences is *true* under that truth assignment. A sentence is *valid* if and only if it is satisfied by *every* truth assignment. A sentence is *satisfiable* if and only if it is satisfied by at least one truth assignment. A sentence is *falsifiable* if and only if there is at least one truth assignment that makes it false. A sentence is *unsatisfiable* if and only if it is not satisfied by any truth assignment. A sentence is *contingent* if and only if it is both satisfiable and falsifiable, i.e., it is neither valid nor unsatisfiable. A set of sentences Δ *logically entails* a sentence φ (written $\Delta \models \varphi$) if and only if every truth assignment that satisfies Δ also satisfies φ.

6.12   EXERCISES

6.1. Say whether each of the following expressions is a syntactically legal sentence of Relational Logic. Assume that *jim* and *molly* are object constants; assume that *person* is a unary relation constant; and assume that *parent* is a binary relation constant.

(a)   *parent(jim, molly)*
(b)   *parent(molly, molly)*
(c)   ¬*person(jim)*
(d)   *person(jim, molly)*
(e)   *parent(molly, z)*
(f)   ∃*x.parent(molly, x)*
(g)   ∃*y.parent(molly, jim)*
(h)   ∀*z.(z(jim, molly)* ⇒ *z(molly, jim))*

6.2.   Consider a language with n object constants and a single binary relation constant.

(a)   How many ground terms are there in this language - n, n^2, 2^n, 2^{n^2}, 2^{2^n}?

(b)   How many ground atomic sentences are there in this language - n, n^2, 2^n, 2^{n^2}, 2^{2^n}?

(c)   How many distinct truth assignments are possible for this language - n, n^2, 2^n, 2^{n^2}, 2^{2^n}?

6.3.   Consider a language with object constants a and b and relation constants p and q where p has arity 1 and q has arity 2. Imagine a truth assignment that makes $p(a)$, $q(a,b)$, $q(b,a)$ true and all other ground atoms false. Say whether each of the following sentences is true or false under this truth assignment.

(a)   ∀*x.(p(x)* ⇒ *q(x,x))*
(b)   ∀*x.*∃*y.q(x,y)*
(c)   ∃*y.*∀*x.q(x,y)*
(d)   ∀*x.(p(x)* ⇒
(e)   ∀*x.p(x)* ⇒ ∃*y.q(y,y)*

6.4.   Consider a state of the Sorority World that satisfies the following sentences.

¬*likes(abby,abby)*	*likes(abby,bess)*	¬*likes(abby,cody)*	*likes(abby,dana)*
likes(bess,abby)	¬*likes(bess,bess)*	*likes(bess,cody)*	¬*likes(bess,dana)*
¬*likes(cody,abby)*	*likes(cody,bess)*	¬*likes(cody,cody)*	*likes(cody,dana)*
likes(dana,abby)	¬*likes(dana,bess)*	*likes(dana,cody)*	¬*likes(dana,dana)*

Say which of the following sentences is satisfied by this state of the world.

(a) *likes(dana,cody)*
(b) ¬*likes(abby,dana)*
(c) *likes(bess,cody)* ∨ *likes(bess,dana)*
(d) ∀*y.(likes(bess,y)* => *likes(abby,y))*
(e) ∀*y.(likes(y,cody)* ⇒ *likes(cody,y))*
(f) ∀*x.*¬*likes(x,x)*

6.5. Consider a version of the Blocks World with just three blocks—*a*, *b*, and *c*. The *on* relation is axiomatized below.

$$\begin{array}{ccc} \neg on(a,a) & on(a,b) & \neg on(a,c) \\ \neg on(b,a) & \neg on(b,b) & on(b,c) \\ \neg on(c,a) & \neg on(c,b) & \neg on(c,c) \end{array}$$

Let's suppose that the *above* relation is defined as follows.

$$\forall x.\forall z.(above(x,z) \Leftrightarrow on(x,z) \vee \exists y.(above(x,y) \wedge above(y,z)))$$

A sentence φ is consistent with a set Δ of sentences if and only if there is a truth assignment that satisfies all of the sentences in $\Delta \cup \{\varphi\}$. Say whether each of the following sentences is consistent with the sentences about *on* and *above* shown above.

(*a*) *above(a,c)*

(*b*) *above(a,a)*

(*c*) *above(c,a)*

6.6. Say whether each of the following sentences is valid, contingent, or unsatisfiable.

(*a*) $\forall x.p(x) \Rightarrow \exists x.p(x)$

(*b*) $\exists x.p(x) \Rightarrow \forall x.p(x)$

(*c*) $\forall x.p(x) \Rightarrow p(x)$

(*d*) $\exists x.p(x) \Rightarrow p(x)$

(*e*) $p(x) \Rightarrow \forall x.p(x)$

(*f*) $p(x) \Rightarrow \exists x.p(x)$

(*g*) $\forall x.\exists y.p(x,y) \Rightarrow \exists y.\forall x.p(x,y)$

(*h*) $\forall x.(p(x) \Rightarrow q(x)) \Rightarrow \exists x.(p(x) \wedge q(x))$

(*i*) $\forall x.(p(x) \Rightarrow q(x)) \wedge \exists x.(p(x) \wedge \neg q(x))$

(*j*) $(\exists x.p(x) \Rightarrow \forall x.q(x)) \vee (\forall x.q(x) \Rightarrow \exists x.r(x))$

6.7. Let Γ be a set of Relational Logic sentences, and let φ and ψ be individual Relational Logic sentences. For each of the following claims, state whether it is true or false.

(*a*) $\forall x.\varphi \models \varphi$

(*b*) $\varphi \models \forall x.\varphi$

(*c*) If $\Gamma \models \neg\varphi[\tau]$ for some ground term τ, then $\Gamma \not\models \forall x.\varphi[x]$

(*d*) If $\Gamma \models \varphi[\tau]$ for some ground term τ, then $\Gamma \models \exists x.\varphi[x]$

(*e*) If $\Gamma \models \varphi[\tau]$ for every ground term τ, then $\Gamma \models \forall x.\varphi[x]$

CHAPTER 7

Relational Analysis

7.1 INTRODUCTION

In Relational Logic, it is possible to analyze the properties of sentences in much the same way as in Propositional Logic. Given a sentence, we can determine its validity, satisfiability, and so forth by looking at possible truth assignments. And we can confirm logical entailment or logical equivalence of sentences by comparing the truth assignments that satisfy them and those that don't.

The main problem in doing this sort of analysis for Relational Logic is that the number of possibilities is even larger than in Propositional Logic. For a language with n object constants and m relation constants of arity k, the Herbrand base has $m * n^k$ elements; and consequently, there are 2^{m*n^k} possible truth assignments to consider. If we have 10 objects and 5 relation constants of arity 2, this means 2^{500} possibilities.

Fortunately, as with Propositional Logic, there are some shortcuts that allow us to analyze sentences in Relational Logic without examining all of these possibilities. In this chapter, we start with the truth table method and then look at some of these more efficient methods.

7.2 TRUTH TABLES

As in Propositional Logic, it is in principle possible to build a truth table for any set of sentences in Relational Logic. This truth table can then be used to determine validity, satisfiability, and so forth or to determine logical entailment and logical equivalence.

As an example, let us assume we have a language with just two object constants a and b and two relation constants p and q. Now consider the sentences shown below, and assume we want to know whether these sentences logically entail $\exists x.q(x)$.

$$p(a) \lor p(b)$$
$$\forall x.(p(x) \Rightarrow q(x))$$

A truth table for this problem is shown below. Each of the first four columns represents one of the elements of the Herbrand base for this language. The two middle columns represent our premises, and the final column represents the conclusion.

$p(a)$	$p(b)$	$q(a)$	$q(b)$	$p(a) \lor p(b)$	$\forall x.(p(x) \Rightarrow q(x))$	$\exists x.q(x)$
1	1	1	1	1	1	1
1	1	1	0	1	0	1
1	1	0	1	1	0	1
1	1	0	0	1	0	0
1	0	1	1	1	1	1
1	0	1	0	1	1	1
1	0	0	1	1	0	1
1	0	0	0	1	0	0
0	1	1	1	1	1	1
0	1	1	0	1	0	1
0	1	0	1	1	1	1
0	1	0	0	1	0	0
0	0	1	1	0	1	1
0	0	1	0	0	1	1
0	0	0	1	0	1	1
0	0	0	0	0	1	0

Looking at the table, we see that there are 12 truth assignments that make the first premise true and nine that make the second premise true and five that make them both true (rows 1, 5, 6, 9, and 11). Note that every truth assignment that makes both premises true also makes the conclusion true. Hence, the premises logically entail the conclusion.

7.3   SEMANTIC TREES

While the Truth Table method works in principle, it is impractical when the tables get very large. As with Propositional Logic, we can sometimes avoid generating such tables by incrementally constructing the corresponding "semantic trees." By interleaving unit propagation and simplification with tree generation, we can often prune away unrewarding subtrees before they are generated and thereby reduce the size of the trees.

7.4   BOOLEAN MODELS

Truth tables and semantic trees are good ways of explicitly representing multiple models for a set of sentences. In some cases, there is just one model.

In this approach, we write out an empty table for each relation and then fill in values based on the constraints of the problem. For example, for any unit constraint, we can immediately enter the corresponding truth value in the appropriate box. Given these partial assignments, we then simplify the constraints (as in the semantic trees method), possibly leading to new unit constraints. We continue until there are no more unit constraints.

As an example, consider the Sorority problem introduced in Chapter 1. We are given the constraints shown below, and we want to know whether Dana likes everyone that Bess likes. In other words, we want to confirm that, in every model that satisfies these sentences, Dana likes everyone that Bess likes.

Dana likes Cody.
Abby does not like Dana.
Dana does not like Abby.
Abby likes everyone that Bess likes.
Bess likes Cody or Dana.
Abby and Dana both dislike Bess.
Cody likes everyone who likes her.
Nobody likes herself.

In this particular case, it turns out that there is just one model that satisfies all of these sentences. The first step in creating this model is to create an empty table for the *likes* relation.

	Abby	Bess	Cody	Dana
Abby				
Bess				
Cody				
Dana				

The data we are given has three units—the fact that Dana likes Cody and the facts that Abby does not like Dana and Dana does not like Abby. Using this information we can refine our model by putting a one into the third box in the fourth row and putting zeros in the fourth box of the first row and the first box of the fourth row.

	Abby	Bess	Cody	Dana
Abby				0
Bess				
Cody				
Dana	0		1	

Now, we know that Abby likes everyone that Bess likes. If Bess likes Dana, then we could conclude that Abby likes Dana as well. We already know that Abby does not like Dana, so Bess must not like Dana either.

	Abby	Bess	Cody	Dana
Abby				0
Bess				0
Cody				
Dana	0		1	

At the same time, we know that Bess likes Cody or Dana. Since Bess does not like Dana, she must like Cody. Once again using the fact that Abby likes everyone whom Bess likes, we know that Abby also likes Cody.

	Abby	Bess	Cody	Dana
Abby			1	0
Bess			1	0
Cody				
Dana	0		1	

Abby and Dana both dislike Bess. Using this fact we can add 0s to the first and last cells of the second column.

	Abby	Bess	Cody	Dana
Abby		0	1	0
Bess			1	0
Cody				
Dana	0	0	1	

On the other hand, Cody likes everyone who likes her. This allows us to put a 1 in every column of the third row where there is a 1 in the corresponding rows of the third column.

	Abby	Bess	Cody	Dana
Abby		0	1	0
Bess			1	0
Cody	1	1		1
Dana	0	0	1	

Since nobody likes herself, we can put a 0 in each cell on the diagonal.

	Abby	Bess	Cody	Dana
Abby	0	0	1	0
Bess		0	1	0
Cody	1	1	0	1
Dana	0	0	1	0

Finally, using the fact that Abby likes everyone that Bess likes, we conclude that Bess does not like Abby. (If she did then Abby would like herself, and we know that that is false.)

	Abby	Bess	Cody	Dana
Abby	0	0	1	0
Bess	0	0	1	0
Cody	1	1	0	1
Dana	0	0	1	0

At this point, we have a complete model, and we can check our conclusion to see that this model satisfies the desired conclusion. In this case, it is easy to see that Dana indeed does like everyone that Bess likes.

We motivated this method by talking about cases where the given sentences have a unique model, as in this case. However, the method can also be of value even when there are multiple possible models. For example, if we had left out the belief that Cody likes everyone who likes her, we would still have eight models (corresponding to the eight possible combinations of feelings Cody has for Abby, Bess, and Dana). Yet, even with this ambiguity, it would be possible to determine whether Dana likes everyone Bess likes using just the portion of the table already filled in.

7.5   NON-BOOLEAN MODELS

As defined in Chapter 6, a model in Relational Logic is an assignment of truth values to the ground atoms of our language. We treat each ground atom in our language as a variable and assign it a single truth value (1 or 0). In general, this is a good way to proceed. However, we can sometimes do better.

Consider, for example, a theory with four object constants and two unary relation constants. In this case, there would be eight elements in the Herbrand base and 2^8 (256) possible truth assignments. Now, suppose we had the constraint that each relation is true of at most a single object. Most of these assignments would not satisfy the single value constraint and thus considering them is wasteful.

Luckily, in cases like this, there is a representation for truth assignments that allows us to eliminate such possibilities and thereby save work. Rather than treating each *ground atom* as a separate variable with its own Boolean value, we can think of each *relation* as a variable with 4 possible values. In order to analyze sentences in such a theory, we would need to consider only 4^2 (16) possibilities.

Even if we search the entire space of assignments, this a significant saving over the pure truth table method. Moreover, we can combine this representation with the techniques described earlier to find assignments for these non-Boolean variables in an even more efficient manner.

The game of Sukoshi illustrates this technique and its benefits. (Sukoshi is similar to Sudoku, but it is smaller and simpler.) A typical Sukoshi puzzle is played on a 4x4 board. In a typical instance of Sukoshi, several of the squares are already filled, as in the example below. The goal of the game is to place the numerals 1 through 4 in the remaining squares of the board in such a way that no numeral is repeated in any row or column.

	4		1
2			
			3
		4	

We can formalize the rules of this puzzle in the language of Logic. Once we have done that, we can use the techniques described here to find a solution.

In our formalization, we use the expression $cell(1,2,3)$ to express the fact that the cell in the first row and the second column contains the numeral 3. For example, we can describe the initial board shown above with the following sentences.

$$cell(1,2,4)$$
$$cell(1,4,1)$$
$$cell(2,1,2)$$
$$cell(3,4,3)$$
$$cell(4,3,4)$$

We use the expression $same(x,y)$ to say that x is the same as y. We can axiomatize $same$ by simply stating when it is true and where it is false. An abbreviated axiomatization is shown below.

$same(1,1)$	$\neg same(2,1)$	$\neg same(3,1)$	$\neg same(4,1)$
$\neg same(1,2)$	$same(2,2)$	$\neg same(3,2)$	$\neg same(4,2)$
$\neg same(1,3)$	$\neg same(2,3)$	$same(3,3)$	$\neg same(4,3)$
$\neg same(1,4)$	$\neg same(2,4)$	$\neg same(3,4)$	$same(3,4)$

Using this vocabulary, we can write the rules defining Sukoshi as shown below. The first sentence expresses the constraint that two cells in that same row can contain the same value. The second sentence expresses the constraint that two cells in that same column can contain the same value. The third constraint expresses the fact that every cell must contain at least one value.

$$\forall x.\forall y.\forall z.\forall w.(\ cell(x,y,w) \land cell(x,z,w) \Rightarrow same(y,z))$$
$$\forall x.\forall y.\forall z.\forall w.(\ cell(x,z,w) \land cell(y,z,w) \Rightarrow same(x,y))$$
$$\forall x.\forall y.\exists w.cell(x,z,w)$$

As a first step in solving this problem, we start by focussing on the fourth column, since two of the cells in that column are already filled. We know that there must be a 4 in one of the cells. It cannot be the first, since that cell contains a 1, and it cannot be the third since that cell contains a 3. It also cannot be the fourth, since there is already a 4 in the third cell of the fourth row. By process of elimination, the 4 must go in the fourth cell of the second row, leading to the board shown below.

	4		1
2			4
			3
		4	

At this point, there is a four in every row and every column except for the first column in the third row. So, we can safely place a four in that cell.

	4		1
2			4
4			3
		4	

Since there is just one empty cell in the fourth column, we know it must be filled with the single remaining value, viz. 2. After adding this value, we have the following board.

	4		1
2			4
4			3
		4	2

Now, let's turn our attention to the first column. We know that there must be a 1 in one of the cells. It cannot be the first, since there is already a 1 in that row, and it cannot be the second or third since those cell already contain values. Consequently, the 1 must go in the first cell of the fourth row.

	4		1
2			4
4			3
1		4	2

Once again, we have a column with all but one cell filled. Column 1 has a 2 in the second cell, a 4 in the third, and a 1 in the fourth. So, we can place a 3 in the first cell of that column.

3	4		1
2			4
4			3
1		4	2

At this point, we can fill in the single empty cell in the first row, leading to the following board.

3	4	2	1
2			4
4			3
1		4	2

And we can fill in the single empty cell in the fourth row as well.

3	4	2	1
2			4
4			3
1	3	4	2

Now, let's consider the second column. We cannot put a 2 in the second cell, since there is already a 2 in that row. Since the first and last cells are already full, the only option is to put the 2 into the third cell.

3	4	2	1
2			4
4	2		3
1	3	4	2

Finishing off the third row leads to the board below.

3	4	2	1
2			4
4	2	1	3
1	3	4	2

Finishing off the third column leads to the following board.

3	4	2	1
2		3	4
4	2	1	3
1	3	4	2

Finally, we can place a 1 in the second cell of the second row. And, with that, the board is full. We have a distinct numeral in every row and every column, as required by the rules.

3	4	2	1
2	1	3	4
4	2	1	3
1	3	4	2

Given the initial assignment in this case, it is fairly easy to find a complete assignment that satisfies the Sukoshi constraints. For other initial assignments, solving the problem is more difficult. However, the techniques described here still work to cut down on the amount of work necessary. In fact, virtually all Sukoshi puzzles can be solved using these techniques without any form of trial and error.

7.6   EXERCISES

7.1. Mr. Red, Mr. White, and Mr. Blue meet for lunch. Each is wearing a red shirt, a white shirt, or a blue shirt. No one is wearing more than one color, and no two are wearing the same color. Mr. Blue tells one of his companions, "Did you notice we are all wearing shirts with different color from our names?", and the other man, who is wearing a white shirt, says, "Wow, that's right!" Use the Boolean model technique to figure out who is wearing what color shirt.

7.2. Amy, Bob, Coe, and Dan are traveling to different places. One goes by train, one by car, one by plane, and one by ship. Amy hates flying. Bob rented his vehicle. Coe tends to get seasick. And Dan loves trains. Use the Boolean models method to figure out which person, used which mode of transportation.

7.3. Sudoku is a puzzle consisting of a 9 × 9 board divided into nine 3 × 3 subboards. In a typical puzzle, several of the squares are already filled, as in the example shown below. The goal of the puzzle is to place the numerals 1 through 9 into the remaining squares of the board in such a way that no numeral is repeated in any row or column or 3 × 3 subboard.

5	8	6					1	2
				5	2	8	6	
2	4		8	1				3
			5		3		9	
				8	1	2	4	
4		5	6			7	3	8
	5		2	3			8	1
7					8			
3	6				5			

Use the techniques described in the Chapter to solve this puzzle.

CHAPTER 8

Relational Proofs

8.1 INTRODUCTION

As with Propositional Logic, we can demonstrate logical entailment in Relational Logic by writing proofs. As with Propositional Logic, it is possible to show that a set of Relational Logic premises logically entails a Relational Logic conclusion if and only if there is a finite proof of the conclusion from the premises. Moreover, it is possible to find such proofs in a finite time.

In this chapter, we start by extending the Fitch system from Propositional Logic to Relational Logic. We then illustrate the system with a few examples. Finally, we talk about soundness and completeness.

8.2 PROOFS

Formal proofs in Relational Logic are analogous to formal proofs in Propositional Logic. The major difference is that there are additional mechanisms to deal with quantified sentences.

The Fitch system for Relational Logic is an extension of the Fitch system for Propositional Logic. In addition to the ten logical rules of inference, there are four ordinary rules of inference for quantified sentences and one additional rule for finite languages. Let's look at each of these in turn. (If you're like me, the prospect of going through a discussion of so many rules of inference sounds a little repetitive and boring. However, it is not so bad. Each of the rules has its own quirks and idiosyncrasies, its own personality. In fact, a couple of the rules suffer from a distinct excess of personality. If we are to use the rules correctly, we need to understand these idiosyncrasies.)

Universal Introduction (UI) allows us to reason from arbitrary sentences to universally quantified versions of those sentences.

Universal Introduction

$$\frac{\varphi}{\forall \nu . \varphi}$$

where ν does not occur free in both φ and an active assumption

Typically, UI is used on sentences with free variables to make their quantification explicit. For example, if we have the sentence *hates(jane,y)*, then, we can infer $\forall y . hates(jane,y)$.

Note that we can also apply the rule to sentences that do not contain the variable that is quantified in the conclusion. For example, from the sentence *hates(jane,jill)*, we can infer $\forall x . hates(jane,jill)$. And, from the sentence *hates(jane,y)*, we can infer $\forall x . hates(jane,y)$. These are not particularly sensible conclusions. However, the results are correct, and the deduction of such results is necessary to ensure that our proof system is complete.

There is one important restriction on the use of Universal Introduction. If the variable being quantified appears in the sentence being quantified, it must not appear free in any *active assumption*, i.e., an assumption in the current subproof or any superproof of that subproof. For example, if there is a subproof with assumption $p(x)$ and from that we have managed to derive $q(x)$, then we cannot just write $\forall x.q(x)$.

If we want to quantify a sentence in this situation, we must first use Implication Introduction to discharge the assumption and then we can apply Universal Introduction. For example, in the case just described, we can first apply Implication Introduction to derive the result $(p(x) \Rightarrow q(x))$ in the parent of the subproof containing our assumption, and we can then apply Universal Introduction to derive $\forall x.(p(x) \Rightarrow q(x))$.

Universal Elimination (UE) allows us to reason from the general to the particular. It states that, whenever we believe a universally quantified sentence, we can infer a version of the target of that sentence in which the universally quantified variable is replaced by an appropriate term.

Universal Elimination

$$\forall v.\varphi[v]$$
$$\overline{\qquad\qquad\qquad\qquad\qquad}$$
$$\varphi[\tau]$$

where τ is substitutable for v in φ

For example, consider the sentence $\forall y.hates(jane,y)$. From this premise, we can infer that Jane hates Jill, i.e., $hates(jane,jill)$. We also can infer that Jane hates her mother, i.e., $hates(jane,mother(jane))$. We can even infer than Jane hates herself, i.e., $hates(jane,jane)$.

In addition, we can use Universal Elimination to create conclusions with free variables. For example, from $\forall x.hates(jane,x)$, we can infer $hates(jane,x)$ or, equivalently, $hates(jane,y)$.

In using Universal Elimination, we have to be careful to avoid conflicts with other variables and quantifiers in the quantified sentence. This is the reason for the constraint on the replacement term. As an example of what can go wrong without this constraint, consider the sentence $\forall x.\exists y.hates(x,y)$, i.e., everybody hates somebody. From this sentence, it makes sense to infer $\exists y.hates(jane,y)$, i.e., Jane hates somebody. However, we do not want to infer $\exists y.hates(y,y)$; i.e., there is someone who hates herself.

We can avoid this problem by obeying the restriction on the Universal Elimination rule. We say that a term τ is *free* for a variable v in a sentence φ if and only if no free occurrence of v occurs within the scope of a quantifier of some variable in τ. For example, the term x is free for y in $\exists z.hates(y,z)$. However, the term z is not free for y, since y is being replaced by z and y occurs within the scope of a quantifier of z. Thus, we cannot substitute z for y in this sentence, and we avoid the problem we have just described.

Existential Introduction (EI) is easy. If we believe a sentence involving a ground term τ, then we can infer an existentially quantified sentence in which one, some, or all occurrences of τ have been replaced by the existentially quantified variable.

Existential Introduction

$$\frac{\varphi[\tau]}{\exists \nu.\varphi[\nu]}$$

For example, from the sentence *hates(jill,jill)*, we can infer that there is someone who hates herself, i.e., $\exists x.hates(x,x)$. We can also infer that there is someone Jill hates, i.e., $\exists x.hates(jill,x)$, and there is someone who hates Jill, i.e., $\exists y.hates(x,jill)$. And, by two applications of Existential Introduction, we can infer that someone hates someone, i.e., $\exists x.\exists y.hates(x,y)$.

Note that, in Existential Introduction, it is important to avoid variables that appear in the sentence being quantified. Without this restriction, starting from $\exists x.hates(jane,x)$, we might deduce $\exists x.\exists x.hates(x,x)$. It is an odd sentence since it contains nested quantifiers of the same variable. However, it is a legal sentence, and it states that there is someone who hates himself, which does not follow from the premise in this case.

Existential Elimination (EE). Suppose we have an existentially quantified sentence with target φ; and suppose we have a universally quantified implication in which the antecedent is the same as the scope of our existentially quantified sentence and the conclusion does not contain any occurrences of the quantified variable. Then, we can use Existential Elimination to infer the consequent.

Existential Elimination

$$\frac{\exists \nu.\varphi[\nu]}{\psi} \\ \forall \nu.(\varphi[\nu] \Rightarrow \psi)$$

where ν does not occur free in ψ

For example, if we have the sentence $\forall x.(hates(jane,x) \Rightarrow \neg nice(jane))$ and we have the sentence $\exists x.hates(jane,x)$, then we can conclude $\neg nice(jane))$.

It is interesting to note that Existential Elimination is analogous to Or Elimination. This is as it should, as an existential sentence is effectively a disjunction. Recall that, in Or Elimination, we start with a disjunction with n disjuncts and n implications, one for each of the disjuncts and produce as conclusion the consequent shared by all of the implications. An existential sentence (like the first premise in any instance of Existential Elimination) is effectively a disjunction over the set of all ground terms; and a universal implication (like the second premise in any instance of Existential Elimination) is effectively a set of implications, one for each ground term in the language. The conclusion of Existential Elimination is the common consequent of these implications, just as in Or Elimination.

Finally, for languages with finite Herbrand bases, we have the *Domain Closure* (DC) rule. For a language with object constants $\sigma_1, \ldots, \sigma_n$, the rule is written as shown below. If we believe a schema is true for every instance, then we can infer a universally quantified version of that schema.

Domain Closure

$\varphi[\sigma_1]$

...

$\dfrac{\varphi[\sigma_n]}{\forall v.\varphi[v]}$

For example, in a language with four object constants a and b and c and d, we can derive the conclusion $\forall x.\varphi[x]$ whenever we have $\varphi[a]$ and $\varphi[b]$ and $\varphi[c]$ and $\varphi[d]$.

Why restrict DC to languages with finitely many ground terms? Why not use domain closure rules for languages with infinitely many ground terms as well? It would be good if we could, but this would require rules of infinite length, and we do not allow infinitely large sentences in our language. We can get the effect of such sentences through the use of *induction*, which is discussed in a later chapter.

As in Chapter 4, we define a *structured proof* of a conclusion from a set of premises to be a sequence of (possibly nested) sentences terminating in an occurrence of the conclusion at the *top level* of the proof. Each step in the proof must be either (1) a premise (at the top level) or an assumption (other than at the top level) or (2) the result of applying an ordinary or structured rule of inference to earlier items in the sequence (subject to the constraints given above and in Chapter 3).

8.3   EXAMPLE

As an illustration of these concepts, consider the following problem. Suppose we believe that everybody loves somebody. And suppose we believe that everyone loves a lover. Our job is to prove that Jack loves Jill.

First, we need to formalize our premises. Everybody loves somebody. For all y, there exists a z such that *loves*(y,z).

$$\forall y.\exists z.loves(y,z)$$

Everybody loves a lover. If a person is a lover, then everyone loves him. If a person loves another person, then everyone loves him. For all x and for all y and for all z, *loves*(y,z) implies *loves*(x,y).

$$\forall x.\forall y.\forall z.(loves(y,z) \Rightarrow loves(x,y))$$

Our goal is to show that Jack loves Jill. In other words, starting with the preceding sentences, we want to derive the following sentence.

$$loves(jack,jill)$$

A proof of this result is shown below. Our premises appear on lines 1 and 2. The sentence on line 3 is the result of applying Universal Elimination to the sentence on line 1, substituting *jill* for y. The sentence on line 4 is the result of applying Universal Elimination to the sentence on line 2, substituting *jack* for x. The sentence on line 5 is the result of applying Universal Elimination to the sentence on line 4, substituting *jill* for y. Finally, we apply Existential Elimination to produce our conclusion on line 6.

1.	$\forall y.\exists z.loves(y,z)$	Premise
2.	$\forall x.\forall y.\forall z.(loves(y,z) \Rightarrow loves(x,y))$	Premise
3.	$\exists z.loves(jill,z)$	UE: 1
4.	$\forall y.\forall z.(loves(y,z) \Rightarrow loves(jack,y))$	UE: 2
5.	$\forall z.(loves(jill,z) \Rightarrow loves(jack,jill))$	UE: 4
6.	$loves(jack,jill)$	EE: 3, 5

Now, let's consider a slightly more interesting version of this problem. We start with the same premises. However, our goal now is to prove the somewhat stronger result that everyone loves everyone. For all x and for all y, x loves y.

$$\forall x.\forall y.loves(x,y)$$

The proof shown below is analogous to the proof above. The only difference is that we have free variables in place of object constants at various points in the proof. Once again, our premises appear on lines 1 and 2. Once again, we use Universal Elimination to produce the result on line 3; but this time the result contains a free variable. We get the results on lines 4 and 5 by successive application of Universal Elimination to the sentence on line 2. We deduce the result on line 6 using Existential Elimination. Finally, we use two applications of Universal Introduction to generalize our result and produce the desired conclusion.

1.	$\forall y.\exists z.loves(y,z)$	Premise
2.	$\forall x.\forall y.\forall z.(loves(y,z) \Rightarrow loves(x,y))$	Premise
3.	$\exists z.loves(y,z)$	UE: 1
4.	$\forall y.\forall z.(loves(y,z) \Rightarrow loves(x,y))$	UE: 2
5.	$\forall z.(loves(y,z) \Rightarrow loves(x,y))$	UE: 4
6.	$loves(x,y)$	EE: 3, 5
7.	$\forall y.loves(x,y)$	UI: 6
8.	$\forall x.\forall y.loves(x,y)$	UI: 7

This second example illustrates the power of free variables. We can manipulate them as though we are talking about specific individuals (though each one could be any object); and, when we are done, we can universalize them to derive universally quantified conclusions.

8.4   EXAMPLE

As another illustration of Relational Logic proofs, consider the following problem. We know that horses are faster than dogs and that there is a greyhound that is faster than every rabbit. We know that Harry is a horse and that Ralph is a rabbit. Our job is to derive the fact that Harry is faster than Ralph.

First, we need to formalize our premises. The relevant sentences follow. Note that we have added two facts about the world not stated explicitly in the problem: that greyhounds are dogs and that our *faster than* relationship is transitive.

$$\forall x.\forall y.(h(x) \land d(y) \Rightarrow f(x,y))$$
$$\exists y.(g(y) \land \forall z.(r(z) \Rightarrow f(y,z)))$$
$$\forall y.(g(y) \Rightarrow d(y))$$
$$\forall x.\forall y.\forall z.(f(x,y) \land f(y,z) \Rightarrow f(x,z))$$
$$h(harry)$$
$$r(ralph)$$

Our goal is to show that Harry is faster than Ralph. In other words, starting with the preceding sentences, we want to derive the following sentence.

$$f(harry,ralph)$$

The derivation of this conclusion goes as shown below. The first six lines correspond to the premises just formalized. On line 7, we start a subproof with an assumption corresponding to the scope of the existential on line 2, with the idea of using Existential Elimination later on in the proof. Lines 8 and 9 come from And Elimination. Line 10 is the result of applying Universal Elimination to the sentence on line 9. On line 11, we use Implication Elimination to infer that y is faster than Ralph. Next, we instantiate the sentence about greyhounds and dogs and infer that y is a dog. Then, we instantiate the sentence about horses and dogs; we use And Introduction to form a conjunction matching the antecedent of this instantiated sentence; and we infer that Harry is faster than y. We then instantiate the transitivity sentence, again form the necessary conjunction, and infer the desired conclusion. Finally, we use Implication Introduction to discharge our subproof; we use Universal Introduction to universalize the results; and we use Existential Elimination to produce our desired conclusion.

1. $\forall x. \forall y. (h(x) \wedge d(y) \Rightarrow f(x,y))$ Premise
2. $\exists y. (g(y) \wedge \forall z. (r(z) \Rightarrow f(y,z)))$ Premise
3. $\forall y. (g(y) \Rightarrow d(y))$ Premise
4. $\forall x. \forall y. \forall z. f(x,y)) \wedge f(y,z) \Rightarrow f(x,z))$ Premise
5. $h(harry)$ Premise
6. $r(ralph)$ Premise
7. | $g(y) \wedge \forall z. (r(z) \Rightarrow f(y,z))$ Assumption
8. | $g(y)$ AE: 7
9. | $\forall z. (r(z) \Rightarrow f(y,z))$ AE: 7
10. | $r(ralph) \Rightarrow f(y,ralph))$ UE: 9
11. | $f(y,ralph))$ IE: 10, 6
12. | $g(y) \Rightarrow d(y)$ UE: 3
13. | $d(y)$ IE: 12, 8
14. | $\forall y. (h(harry) \wedge d(y) \Rightarrow f(harry,y))$ UE: 1
15. | $h(harry) \wedge d(y) \Rightarrow f(harry,y)$ UE: 14
16. | $h(harry) \wedge d(y)$ AI: 5. 13
17. | $f(harry,y)$ IE: 15, 16
18. | $\forall y. \forall z. (f(harry,y) \wedge f(y,z) \Rightarrow f(harry,z))$ UE: 4
19. | $\forall z. (f(harry,y) \wedge f(y,z) \Rightarrow f(harry,z))$ UE: 18
20. | $f(harry,y) \wedge f(y,ralph) \Rightarrow f(harry,ralph)$ UE: 19
21. | $f(harry,y) \wedge f(y,ralph)$ AI: 17, 11
22. | $f(harry,ralph)$ IE: 20, 21
23. $g(y) \wedge \forall z. (r(z) \Rightarrow f(y,z)) \Rightarrow f(harry,ralph)$ II: 7, 22
24. $\forall y. (g(y) \wedge \forall z. (r(z) \Rightarrow f(y,z)) \Rightarrow f(harry,ralph))$ UI: 23
25. $f(harry,ralph)$ EE: 2, 24

This derivation is somewhat lengthy, but it is completely mechanical. Each conclusion follows from previous conclusions by a mechanical application of a rule of inference. On the other hand, in producing this derivation, we rejected numerous alternative inferences. Making these choices intelligently is one of the key problems in the process of inference.

8.5   EXAMPLE

In this section, we use our proof system to prove some basic results involving quantifiers.

Given $\forall x.\forall y.(p(x,y) \Rightarrow q(x))$, we know that $\forall x.(\exists y.p(x,y) \Rightarrow q(x))$. In general, given a universally quantified implication, it is okay to drop a universal quantifier of a variable outside the implication and apply an existential quantifier of that variable to the antecedent of the implication, provided that the variable does not occur in the consequent of the implication.

Our proof is shown below. As usual, we start with our premise. We start a subproof with an existential sentence as assumption. Then, we use Universal Elimination to strip away the outer quantifier from the premise. This allows us to derive $q(x)$ using Existential Elimination. Finally, we create an implication with Implication Introduction, and we generalize using Universal Introduction.

1.	$\forall x.\forall y.(p(x,y) \Rightarrow q(x))$	Premise
2.	$\exists y.p(x,y)$	Assumption
3.	$\forall y.(p(x,y) \Rightarrow q(x))$	UE: 1
4.	$q(x)$	EE: 2, 3
5.	$\exists y.p(x,y) \Rightarrow q(x)$	II: 4
6.	$\forall x(\exists y.p(x,y) \Rightarrow q(x))$	UI: 5

The relationship holds the other way around as well. Given $\forall x.(\exists y.p(x,y) \Rightarrow q(x))$, we know that $\forall x.\forall y.(p(x,y) \Rightarrow q(x))$. We can convert an existential quantifier in the antecedent of an implication into a universal quantifier outside the implication.

Our proof is shown below. As usual, we start with our premise. We start a subproof by making an assumption. Then we turn the assumption into an existential sentence to match the antecedent of the premise. We use Universal Implication to strip away the quantifier in the premise to expose the implication. Then, we apply Implication Elimination to derive $q(x)$. Finally, we create an implication with Implication Introduction, and we generalize using two applications of Universal Introduction.

1.	$\forall x.(\exists y.(p(x,y) \Rightarrow q(x))$	Premise
2.	$p(x,y)$	Assumption
3.	$\exists y.p(x,y)$	EI: 2
4.	$\exists y.p(x,y) \Rightarrow q(x)$	UE: 1
5.	$q(x)$	IE: 4, 3
6.	$p(x,y) \Rightarrow q(x)$	II: 5
7.	$\forall x.\forall y.(p(x,y) \Rightarrow q(x))$	$2 \times$ UI: 6

RECAP

A Fitch system for Relational Logic can be obtained by extending the Fitch system for Propositional Logic with four additional rules to deal with quantifiers. The *Universal Introduction* rule allows us to reason from arbitrary sentences to universally quantified versions of those sentences. The *Universal Elimination* rule allows us to reason from a universally quantified sentence to a version of the target of that sentence in which the universally quantified variable is replaced by an appropriate term. The *Existential Introduction* rule allows us to reason from a sentence involving a term τ to an existentially quantified sentence in which one, some, or all occurrences of τ have been replaced by the existentially quantified variable. Finally, the *Existential Elimination* rule allows us to reason from an existentially quantified sentence $\exists v.\varphi[v]$ and a universally quantified implication $\forall v.(\varphi[v] \Rightarrow \psi)$ to the consequent ψ, under the condition that v does not occur in ψ.

8.6   EXERCISES

8.1. Given $\forall x.(p(x) \wedge q(x))$, use the Fitch System to prove $\forall x.p(x) \wedge \forall x.q(x)$.

8.2. Given $\forall x.(p(x) \Rightarrow q(x))$, use the Fitch System to prove $\forall x.p(x) \Rightarrow \forall x.q(x)$.

8.3. Given the premises $\forall x.(p(x) \Rightarrow q(x))$ and $\forall x.(q(x) \Rightarrow r(x))$, use the Fitch system to prove the conclusion $\forall x.(p(x) \Rightarrow r(x))$.

8.4. Given $\forall x.\forall y.p(x,y)$, use the Fitch System to prove $\forall y.\forall x.p(x,y)$.

8.5. Given $\forall x.\forall y.p(x,y)$, use the Fitch System to prove $\forall x.\forall y.p(y,x)$.

8.6. Given $\exists y.\forall x.p(x,y)$, use the Fitch system to prove $\forall x.\exists y.p(x,y)$.

8.7. Given $\exists x.\neg p(x)$, use the Fitch System to prove $\neg\forall x.p(x)$.

8.8. Given $\forall x.p(x)$, use the Fitch System to prove $\neg\exists x.\neg p(x)$.

C H A P T E R   9

Herbrand Logic

9.1   INTRODUCTION

Relational Logic, as defined in Chapter 6, allows us to axiomatize worlds with varying numbers of objects. The main restriction is that the worlds must be finite (since we have only finitely many constants to refer to these objects).

Often, we want to describe worlds with infinitely many objects. For example, it would be nice to axiomatize arithmetic over the integers or to talk about sequences of objects of varying lengths. Unfortunately, this is not possible due to the finiteness restriction of Relational Logic.

One way to get infinitely many terms is to allow our vocabulary to have infinitely many object constants. While there is nothing wrong with this in principle, it makes the job of axiomatizing things effectively impossible, as we would have to write out infinitely many sentences in many cases.

In this chapter, we explore an alternative to Relational Logic, called Herbrand Logic, in which we can name infinitely many objects with a finite vocabulary. The trick is to expand our language to include not just object constants but also complex terms that can be built from object constants in infinitely many ways. By constructing terms in this way, we can get infinitely many names for objects; and, because our vocabulary is still finite, we can finitely axiomatize some things in a way that would not be possible with infinitely many object constants.

In this chapter, we proceed through the same stages as in the introduction to Relational Logic. We start with syntax and semantics. We then discuss evaluation and satisfaction. We look at some examples. And we conclude with a discussion of some of the properties of Herbrand Logic.

9.2   SYNTAX AND SEMANTICS

The syntax of Herbrand Logic is the same as that of Relational Logic except for the addition of *function constants* and *functional expressions*.

As we shall see, function constants are similar to relation constants in that they are used in forming complex expressions by combining them with an appropriate number of arguments. Accordingly, each function constant has an associated *arity*, i.e., the number of arguments with which that function constant can be combined. A function constant that can combined with a single argument is said to be *unary*; one that can be combined with two arguments is said to be *binary*; one that can be combined with three arguments is said to be *ternary*; more generally, a function constant that can be combined with n arguments is said to be n-ary.

A *functional expression*, or *functional term*, is an expression formed from an n-ary function constant and n terms enclosed in parentheses and separated by commas. For example, if f is a binary function constant and if a and y are terms, then $f(a,y)$ is a functional expression, as are $f(a,a)$ and $f(y,y)$.

Note that, unlike relational sentences, functional expressions can be nested within other functional expressions. For example, if g is a unary function constant and if a is a term, $g(a)$ and $g(g(a))$ and $g(g(g(a)))$ are all functional expressions.

Finally, in Herbrand Logic, we define a *term* to be a variable or an object constant or a functional expression. The definition here is the same as before except for the addition of functional expressions to this list of possibilities.

And that is all. Relational sentences, logical sentences, and quantified sentences are defined exactly as for ordinary Relational Logic. The only difference between the two languages is the Herbrand Logic allows for functional expressions inside of sentences whereas ordinary Relational Logic does not.

The semantics of Herbrand Logic is effectively the same as that of Relational Logic. The key difference is that, in the presence of functions, the Herbrand base for such a language is infinitely large.

As before, we define the *Herbrand base* for a vocabulary to be the set of all ground relational sentences that can be formed from the constants of the language. Said another way, it is the set of all sentences of the form $r(t_1,...,t_n)$, where r is an n-ary relation constant and $t_1, ... , t_n$ are ground terms.

For a vocabulary with a single object constant a and a single unary function constant f and a single unary relation constant r, the Herbrand base consists of the sentences shown below.

$$\{r(a), r(f(a)), r(f(f(a))), ...\}$$

A *truth assignment* for Herbrand Logic is a mapping that gives each ground relational sentence in the Herbrand base a unique truth value. This is the same as for Relational Logic. The main difference from Relational Logic is that, in this case, a truth assignment is necessarily infinite, since there are infinitely many elements in the Herbrand Base.

Luckily, things are not always so bad. In some cases, only finitely many elements of the Herbrand base are true. In such cases, we can describe a truth assignment in finite space by writing out the elements that are true and assuming that all other elements are false. We shall see some examples of this in the coming sections.

The rules defining the truth of logical sentences in Herbrand Logic are the same as those for logical sentences in Propositional Logic and Relational Logic, and the rules for quantified sentences in Herbrand Logic are exactly the same as those for Relational Logic.

9.3   EVALUATION AND SATISFACTION

The concept of evaluation for Herbrand Logic is the same as that for Relational Logic. Unfortunately, evaluation is usually not practical in this case for two reasons. First of all, truth assignments are infinite in size and so we cannot always write them down. Even when we can finitely characterize a truth assignment (e.g., when the set of true sentences is finite), we may not be able to evaluate quantified sentences mechanically in all cases. In the case of a universally quantified formula, we need to check all instances of the scope, and there are infinitely many possibilities. In the case of an existentially quantified sentence, we need to enumerate possibilities until we find one that succeeds, and we may never find one if the existentially quantified sentence is false.

Satisfaction has similar difficulties. The truth tables for Herbrand Logic are infinitely large and so we cannot write out or check all possibilities.

The good news is that, even though evaluation and satisfaction are not directly computable, there are effective procedures for indirectly determining validity, contingency, unsatisfiability, logical entailment, and so forth that work in many cases even when our usual direct methods fail. The key is symbolic manipulation of various sorts, e.g., the generation of proofs, which we describe in the next few chapters. But, first, in order to gain some intuitions about the power of Herbrand Logic, we look at some examples.

9.4   EXAMPLE–PEANO ARITHMETIC

Peano Arithmetic differs from Modular Arithmetic (axiomatized in Section 6.8) in that it applies to all natural numbers (0, 1, 2, 3, ...). Since there are infinitely many such numbers, we need infinitely many terms.

As mentioned in the introduction, we can get infinitely many terms by expanding our vocabulary with infinitely many object constants. Unfortunately, this makes the job of axiomatizing arithmetic effectively impossible, as we would have to write out infinitely many sentences.

An alternative approach is to represent numbers using a single object constant (e.g., 0) and a single unary function constant (e.g., s). We can then represent every number n by applying the function constant to 0 exactly n times. In this encoding, $s(0)$ represents 1; $s(s(0))$ represents 2; and so forth. With this encoding, we automatically get an infinite universe of terms, and we can write axioms defining addition and multiplication as simple variations on the axioms of Modular Arithmetic.

Unfortunately, even with this representation, axiomatizing Peano Arithmetic is more challenging than axiomatizing Modular Arithmetic. We cannot just write out ground relational sentences to characterize our relations, because there are infinitely many cases to consider. For Peano Arithmetic, we must rely on logical sentences and quantified sentences, not just because they are more economical but because they are the only way we can characterize our relations in finite space.

Let's look at the same relation first. The axioms shown here define the *same* relation in terms of 0 and *s*.

$$\forall x.same(x,x)$$
$$\forall x.(\neg same(0,s(x)) \wedge \neg same(s(x),0))$$
$$\forall x.\forall y.(\neg same(x,y) \Rightarrow \neg same(s(x),s(y)))$$

It is easy to see that these axioms completely characterize the *same* relation. By the first axiom, the same relation holds of every term and itself.

$$same(0,0)$$
$$same(s(0),s(0))$$
$$same(s(s(0)),s(s(0)))$$

...

The other two axioms tell us what is not true. The second axiom tells us that 0 is not the same as any composite term. The same holds true with the arguments reversed.

$\neg same(0,s(0))$	$\neg same(s(0),0)$
$\neg same(0,s(s(0)))$	$\neg same(s(s(0)),0)$
$\neg same(0,s(s(s(0))))$	$\neg same(s(s(s(0))),0)$
...	...

The third axiom builds on these results to show that non-identical composite terms of arbitrary complexity do not satisfy the same relation. Viewed the other way around, to see that two non-identical terms are not the same, we just strip away occurrences of *s* from each term till one of the two terms becomes 0 and the other one is not 0. By the second axiom, these are not the same, and so the original terms are not the same.

$\neg same(s(0),s(s(0)))$	$\neg same(s(s(0)),s(0))$
$\neg same(s(0),s(s(s(0))))$	$\neg same(s(s(s(0))),s(0))$
$\neg same(s(0),s(s(s(s(0)))))$	$\neg same(s(s(s(s(0)))),s(0))$
...	...

Once we have the *same* relation, we can define the other relations in our arithmetic. The following axioms define the plus relation in terms of 0, *s*, and *same*. Adding 0 to any number results in that number. If adding a number *x* to a number *y* produces a number *z*, then adding the successor of *x* to *y* produces the successor of *z*. Finally, we have a functionality axiom for *plus*.

$$\forall y.plus(0,y,y)$$
$$\forall x.\forall y.\forall z.(plus(x,y,z) \Rightarrow plus(s(x),y,s(z)))$$
$$\forall x.\forall y.\forall z.\forall w.(\, plus(x,y,z) \wedge \neg same(z,w) \Rightarrow \neg plus(x,y,w))$$

The axiomatization of multiplication is analogous. Multiplying any number by 0 produces 0. If a number *z* is the product of *x* and *y* and *w* is the sum of *y* and *z*, then *w* is the product of the successor of *x* and *y*. As before, we have a functionality axiom.

$$\forall y.times(0,y,0)$$
$$\forall x.\forall y.\forall z.\forall w.(\ times(x,y,z) \wedge plus(y,z,w) \Rightarrow times(s(x),y,w))$$
$$\forall x.\forall y.\forall z.\forall w.(\ times(x,y,z) \wedge \neg same(z,w) \Rightarrow \neg times(x,y,w))$$

That's all we need—just three axioms for *same* and three axioms for each arithmetic function.

Before we leave our discussion of Peano arithmetic, it is worthwhile to look at the concept of Diophantine equations. A *polynomial equation* is a sentence composed using only addition, multiplication, and exponentiation with fixed exponents (that is numbers not variables). For example, the expression shown below in traditional math notation is a polynomial equation.

$$x^2 + 2y = 4z$$

A *natural Diophantine equation* is a polynomial equation in which the variables are restricted to the natural numbers. For example, the polynomial equation here is also a Diophantine equation and happens to have a solution in the natural numbers, viz. $x = 4$ and $y = 8$ and $z = 8$.

Diophantine equations can be readily expressed as sentences in Peano Arithmetic. For example, we can represent the Diophantine equation above with the sentence shown below.

$$\exists x.\exists y.\exists z.\forall u.\forall v.\forall w.(times(x,x,u) \wedge times(2,y,v) \wedge plus(u,v,w) \Rightarrow times(4,z,w))$$

This is a little messy, but it is doable. And we can always clean things up by adding a little syntactic sugar to our notation to make it look like traditional math notation.

Once this mapping is done, we can use the tools of logic to work with these sentences. In some cases, we can find solutions; and, in some cases, we can prove that solutions do not exist. This has practical value in some situations, but it also has significant theoretical value in establishing important properties of Relational Logic, a topic that we discuss in a later section.

9.5   EXAMPLE–LINKED LISTS

A list is finite sequence of objects. Lists can be flat, e.g., $[a, b, c]$. Lists can also be nested within other lists, e.g., $[a, [b, c], d]$.

A linked list is a way of representing nested lists of variable length and depth. Each element is represented by a cell containing a value and a pointer to the remainder of the list. Our goal in this example is to formalize linked lists and define some useful relations.

To talk about lists of arbitrary length and depth, we use the binary function constant *cons*, and we use the object constant *nil* to refer to the empty list. In particular, a term of the form $cons(\tau_1, \tau_2)$ designates a sequence in which τ_1 denotes the first element and τ_2 denotes the rest of the list. With this function constant, we can encode the list $[a, b, c]$ as follows.

$$cons(a, cons(b, cons(c, nil)))$$

The advantage of this representation is that it allows us to describe functions and relations on lists without regard to length or depth.

As an example, consider the definition of the binary relation *member*, which holds of an object and a list if the object is a top-level member of the list. Using the function constant *cons*, we can characterize the *member* relation as shown below. Obviously, an object is a member of a list if it is the first element; however, it is also a member if it is member of the rest of the list.

$$\forall x.\forall y.member(x, cons(x, y))$$
$$\forall x.\forall y.\forall z.(member(x, z) \Rightarrow member(x, cons(y, z)))$$

In similar fashion, we can define functions to manipulate lists in different ways. For example, the following axioms define a relation called *append*. The value of *append* (its last argument) is a list consisting of the elements in the list supplied as its first argument followed by the elements in the list supplied as its second. For example, we would have *append(cons(a,nil), cons(b, cons(c, nil)), cons(a, cons(b, cons(c, nil))))*. And, of course, we need negative axioms as well (omitted here).

$$\forall z.append(nil, z, z)$$
$$\forall x.\forall y.\forall z.(append(y, z, w) \Rightarrow append(cons(x, y), z, cons(x,w)))$$

We can also define relations that depend on the structure of the elements of a list. For example, the *among* relation is true of an object and a list if the object is a member of the list, if it is a member of a list that is itself a member of the list, and so on. (And, once again, we need negative axioms.)

$$\forall x.among(x, x)$$
$$\forall x.\forall y.\forall z.(among(x, y) \lor among(x, z) \Rightarrow among(x, cons(y, z)))$$

Lists are an extremely versatile representational device, and the reader is encouraged to become as familiar as possible with the techniques of writing definitions for functions and relations on lists. As is true of many tasks, practice is the best approach to gaining skill.

9.6 EXAMPLE–PSEUDO ENGLISH

Pseudo English is a formal language that is intended to approximate the syntax of the English language. One way to define the syntax of Pseudo English is to write grammatical rules in Backus-Naur Form (BNF). The rules shown below illustrate this approach for a small subset of Pseudo English. A sentence is a noun phrase followed by a verb phrase. A noun phrase is either a noun or two nouns separated by the word *and*. A verb phrase is a verb followed by a noun phrase. A noun is either the word *Mary* or the word *Pat* or the word *Quincy*. A verb is either *like* or *likes*.

```
<sentence> ::= <np> <vp>
<np> ::= <noun>
<np> ::= <noun> "and" <noun>
<vp> ::= <verb> <np>
<noun> ::= "mary" | "pat" | "quincy"
<verb> ::= "like" | "likes"
```

Alternatively, we can use Herbrand Logic to formalize the syntax of Pseudo English. The sentences shown below express the grammar described in the BNF rules above. (We have dropped the universal quantifiers here to make the rules a little more readable.) Here, we are using the *append* relation defined in the section of lists.

$np(x) \land vp(y) \land append(x,y,z) \Rightarrow sentence(z)$
$noun(x) \Rightarrow np(x)$
$noun(x) \land noun(y) \land append(x,\text{and},z) \land append(z,y,w) \Rightarrow np(w)$
$verb(x) \land np(y) \land append(x,y,z) \Rightarrow vp(z)$
$noun(mary)$
$noun(pat)$
$noun(quincy)$
$verb(like)$
$verb(likes)$

Using these sentences, we can test whether a given sequence of words is a syntactically legal sentence in Pseudo English and we can use our logical entailment procedures to enumerate syntactically legal sentences, like those shown below.

mary likes pat
pat and quincy like mary
mary likes pat and quincy

One weakness of our BNF and the corresponding axiomatization is that there is no concern for agreement in number between subjects and verbs. Hence, with these rules, we can get the following expressions, which in Natural English are ungrammatical.

× *mary like pat*
× *pat and quincy likes mary*

Fortunately, we can fix this problem by elaborating our rules just a bit. In particular, we add an argument to some of our relations to indicate whether the expression is singular or plural. Here, 0 means singular, and 1 means plural. We then use this to block sequences of words where the numbers do not agree.

$np(x,w) \land vp(y,w) \land append(x,y,z) \Rightarrow sentence(z)$
$noun(x) \Rightarrow np(x,0)$
$noun(x) \land noun(y) \land append(x,\text{and},z) \land append(z,y,w) \Rightarrow np(w,1)$
$verb(x,w) \land np(y,v) \land append(x,y,z) \Rightarrow vp(z,w)$
$noun(mary)$
$noun(pat)$
$noun(quincy)$
$verb(like,1)$
$verb(likes,0)$

With these rules, the syntactically correct sentences shown above are still guaranteed to be sentences, but the ungrammatical sequences are blocked. Other grammatical features can be formalized in similar fashion, e.g., gender agreement in pronouns (*he* vs. *she*), possessive adjectives (*his* vs. *her*), reflexives (like *himself* and *herself*), and so forth.

9.7   EXAMPLE–METALEVEL LOGIC

Throughout this book, we have been writing sentences in English about sentences in Logic, and we have been writing informal proofs in English about formal proofs in Logic. A natural question to ask is whether it is possible formalize Logic within Logic. The answer is yes. The limits of what can be done are very interesting. In this section, we look at a small subset of this problem, viz. using Relational Logic to formalize information about Propositional Logic.

The first step in formalizing Propositional Logic in Relational Logic is to represent the syntactic components of Propositional Logic.

In what follows, we make each proposition constant in our Propositional Logic language an object constant in our Relational Logic formalization. For example, if our Propositional Logic language has relation constants p, q, and r, then p, q, and r are object constants in our formalization.

Next, we introduce function constants to represent constructors of complex sentences. There is one function constant for each logical operator—*not* for $\neg$, *and* for $\wedge$, *or* for $\vee$, *if* for $\Rightarrow$, and *iff* for $\Leftrightarrow$. Using these function constants, we represent Propositional Logic sentences as terms in our language. For example, we use the term *and*(p,q) to represent the Propositional Logic sentence ($p \wedge q$); and we use the term *if*(*and*(p,q),r) to represent the Propositional Logic sentence ($p \wedge q \Rightarrow r$).

Finally, we introduce a selection of relation constants to express the types of various expressions in our Propositional Logic language. We use the unary relation constant *proposition* to assert that an expression is a proposition. We use the unary relation constant *negation* to assert that an expression is a negation. We use the unary relation constant *conjunction* to assert that an expression is a conjunction. We use the unary relation constant *disjunction* to assert that an expression is a disjunction. We use the unary relation constant *implication* to assert that an expression is an implication. We use the unary relation constant *biconditional* to assert that an expression is a biconditional. And we use the unary relation constant *sentence* to assert that an expression is a proposition.

With this vocabulary, we can characterize the syntax of our language as follows. We start with declarations of our proposition constants.

$$proposition(p)$$
$$proposition(q)$$
$$proposition(r)$$

Next, we define the types of expressions involving our various logical operators.

$$\forall x.(sentence(x) \Rightarrow negation(not(x)))$$
$$\forall x.\forall y.(sentence(x) \wedge sentence(y) \Rightarrow conjunction(and(x,y)))$$
$$\forall x.\forall y.(sentence(x) \wedge sentence(y) \Rightarrow disjunction(or(x,y)))$$
$$\forall x.\forall y.(sentence(x) \wedge sentence(y) \Rightarrow implication(if(x,y)))$$
$$\forall x.\forall y.(sentence(x) \wedge sentence(y) \Rightarrow biconditional(iff(x,y)))$$

Finally, we define sentences as expressions of these types.

$$\forall x.(proposition(x) \Rightarrow sentence(x))$$
$$\forall x.(negation(x) \Rightarrow sentence(x))$$
$$\forall x.(conjunction(x) \Rightarrow sentence(x))$$
$$\forall x.(disjunction(x) \Rightarrow sentence(x))$$
$$\forall x.(implication(x) \Rightarrow sentence(x))$$
$$\forall x.(biconditional(x) \Rightarrow sentence(x))$$

Note that these sentences constrain the types of various expressions but do not define them completely. For example, we have not said that $not(p)$ is *not* a conjunction. It is possible to make our definitions more complete by writing negative sentences. However, they are a little messy, and we do not need them for the purposes of this section.

With a solid characterization of syntax, we can formalize our rules of inference. We start by representing each rule of inference as a relation constant. For example, we use the ternary relation constant *ai* to represent And Introduction, and we use the binary relation constant *ae* to represent And Elimination. With this vocabulary, we can define these relations as shown below.

$$\forall x.\forall y.(sentence(x) \wedge sentence(y) \Rightarrow ai(x,y,and(x,y)))$$
$$\forall x.\forall y.(sentence(x) \wedge sentence(y) \Rightarrow ae(and(x,y),x))$$
$$\forall x.\forall y.(sentence(x) \wedge sentence(y) \Rightarrow ae(and(x,y),y))$$

In similar fashion, we can define proofs—both linear and structured. We can even define truth assignments, satisfaction, and the properties of validity, satisfiability, and so forth. Having done all of this, we can use the proof methods discussed in the next chapters to prove our metatheorems about Propositional Logic.

We can use a similar approach to formalizing Relational Logic within Relational Logic. However, in that case, we need to be very careful. If done incorrectly, we can write paradoxical sentences, i.e., sentences that are neither true nor false. For example, a careless formalization leads to formal versions of sentences like *This sentence is false*, which is self-contradictory, i.e., it cannot be true and cannot be false. Fortunately, with care it is possible to avoid such paradoxes and thereby get useful work done.

9.8   UNDECIDABILITY

The good news about Herbrand Logic is that it is highly expressive. We can formalize things in Herbrand Logic that cannot be formalized (at least in finite form) in Relational Logic. For example, we showed how to define addition and multiplication in finite form. This is not possible with Relational Logic and in other logics (e.g., First-Order Logic).

The bad news is that the questions of unsatisfiability and logical entailment for Herbrand Logic are not effectively computable. Explaining this in detail is beyond the scope of this course. However, we can give a line of argument that suggests why it is true. The argument reduces a problem that is generally accepted to be non-semidecidable to the question of unsatisfiability / logical entailment for Herbrand Logic. If our logic were semidecidable, then this other question would be semidecidable as well; and, since it is known not to be semidecidable, then Herbrand Logic must not be semidecidable either.

As we know, Diophantine equations can be readily expressed as sentences in Herbrand Logic. For example, we can represent the solvability of Diophantine equation $3x^2 = 1$ with the sentence shown below.

$$\exists x.\exists y.(\mathit{times}(x, x, y) \land \mathit{times}(s(s(s(0))), y, s(0)))$$

We can represent every Diophantine equation in an analogous way. We can express the unsolvability of a Diophantine equation by negating the corresponding sentence. We can then ask the question of whether the axioms of arithmetic logically entail this negation or, equivalently, whether the axioms of Arithmetic together with the unnegated sentence are unsatisfiable.

The problem is that it is well known that determining whether Diophantine equations are unsolvable is not semidecidable. If we could determine the unsatisfiability of our encoding of a Diophantine equation, we could decide whether it is unsolvable, contradicting the non-semidecidability of that problem.

Note that this does not mean Herbrand Logic is useless. In fact, it is great for expressing such information; and we can prove many results, even though, in general, we cannot prove everything that follows from arbitrary sets of sentences in Herbrand Logic. We discuss this issue further in later chapters.

RECAP

Herbrand Logic is an extended version of Relational Logic that includes *functional expressions*. Since functional expressions can be composed with each other in infinitely many ways, the Herbrand base for Herbrand Logic is infinite, allowing us to axiomatize infinite relations with a finite vocabulary. Other than the addition of functional expressions, the syntax and semantics of Herbrand Logic is the same as that of Relational Logic. Questions of unsatisfiability and logical entailment can sometimes be computed in Herbrand Logic, though in general those questions are not effectively computable.

9.9   EXERCISES

9.1.   Say whether each of the following expressions is a syntactically legal sentence of Herbrand Logic. Assume that a and b are object constants, f is a unary function constant, and p is a unary relation constant.

  (a) $p(a)$
  (b) $p(f(a))$
  (c) $f(f(a))$
  (d) $p(f(f(a)))$
  (e) $p(f(p(a)))$

9.2.   Say whether each of the following sentences is logically entailed by the sentences in Section 9.4.

  (a) $same(s(0),s(s(s(0))))$
  (b) $plus(s(s(0)),s(s(s(0))),s(s(s(s(s(0))))))$
  (c) $times(s(s(0)),s(s(s(0))),s(s(s(s(s(s(0)))))))$
  (d) $times(s(0),s(s(s(0))),s(s(s(0))))$

9.3.   Say whether each of the following sentences is logically entailed by the sentences in Section 9.5.

  (a) $append(nil, nil, nil)$
  (b) $append(cons(a, nil), nil, cons(a, nil))$
  (c) $append(cons(a, nil), cons(b, nil), cons(a, b))$
  (d) $append(cons(cons(a, nil), nil), cons(b, nil), cons(a, cons(b, nil)))$

9.4.   Say whether each of the following sentences is a grammatical sentence of Pseudo English according to the enhanced grammar presented at the end of Section 9.6.

  (a)   *Mary likes Pat and Quincy.*
  (b)   *Mary likes Pat and Mary likes Quincy.*
  (c)   *Mary likes Mary.*
  (d)   *Mary likes herself.*

9.5.   Say whether each of the following sentences is logically entailed by the sentences in Section 9.7.

  (a) $conjunction(and(not(p), not(q)))$
  (b) $conjunction(not(or(not(p), not(q))))$
  (c) $ae(and(p, or(p, q)), or(p, q))$
  (d) $ae(and(p, or(p, q)), and(p, q))$

CHAPTER 10

Herbrand Proofs

10.1 INTRODUCTION

Logical entailment for Herbrand Logic is defined the same as for Propositional Logic and Relational Logic. A set of premises logically entails a conclusion if and only if every truth assignment that satisfies the premises also satisfies the conclusions. In the case of Propositional Logic and Relational Logic, we can check logical entailment by examining a truth table for the language. With finitely many proposition constants, the truth table is large but finite. For Herbrand Logic, things are not so easy. It is possible to have Herbrand bases of infinite size; and, in such cases, truth assignments are infinitely large and there are infinitely many of them, making it impossible to check logical entailment using truth tables.

All is not lost. As with Propositional Logic and Relational Logic, we can establish logical entailment in Herbrand Logic by writing proofs. In fact, it is possible to show that, with a few simple restrictions, a set of premises logically entails a conclusions if and only if there is a finite proof of the conclusion from the premises, even when the Herbrand base is infinite. Moreover, it is possible to find such proofs in a finite time. That said, things are not perfect. If a set of sentences does *not* logically entail a conclusion, then the process of searching for a proof might go on forever. Moreover, if we remove the restrictions mentioned above, we lose the guarantee of finite proofs. Still, the relationship between logical entailment and finite provability, given those restrictions, is a very powerful result and has enormous practical benefits.

In this chapter, we talk about the non-compactness of Herbrand Logic and the loss of completeness in our proof procedure. In the next chapter, we look at an extension to Fitch, called Induction, that allows us to prove more results in Herbrand Logic.

10.2 NON-COMPACTNESS AND INCOMPLETENESS

In light of the negative results above, namely that Herbrand Logic is inherently incomplete, it is not surprising that Herbrand Logic is not compact. Recall that compactness says that, if an infinite set of sentences is unsatisfiable, then there is some finite subset that is unsatisfiable. It guarantees finite proofs.

Non-Compactness Theorem: Herbrand Logic is not compact.

Proof. Consider the following infinite set of sentences.

$$p(a), p(f(a)), p(f(f(a))), ...$$

Assume the vocabulary is $\{p, a, f\}$. Hence, the ground terms are a, $f(a)$, $f(f(a))$, ... This set of sentences entails $\forall x.p(x)$. Add in the sentence $\exists x.\neg p(x)$. Clearly, this infinite set is unsatisfiable. However, every finite subset is satisfiable. (Every finite subset is missing either $\exists x.\neg p(x)$ or one of the sentences above. If it is the former, the set is satisfiable; and, if it is the latter, the set can be satisfied by making the missing sentence false.) Thus, compactness does not hold.

Corollary (Infinite Proofs): In Herbrand Logic, some entailed sentences have only infinite proofs.

Proof. The above proof demonstrates a set of sentences that entail $\forall x.p(x)$. The set of premises in any finite proof will be missing one of the above sentences; thus, those premises do not entail $\forall x.p(x)$. Thus no finite proof can exist for $\forall x.p(x)$.

The statement in this Corollary was made earlier with the condition that checking whether a candidate proof actually proves a conjecture is decidable. There is no such condition on this theorem.

CHAPTER 11

Induction

11.1 INTRODUCTION

Induction is reasoning from the specific to the general. If various instances of a schema are true and there are no counterexamples, we are tempted to conclude a universally quantified version of the schema.

$$
\begin{aligned}
p(a) &\Rightarrow q(a) \\
p(b) &\Rightarrow q(b) \quad \rightarrow \quad \forall x.p(x) \Rightarrow q(x) \\
p(c) &\Rightarrow q(c)
\end{aligned}
$$

Incomplete induction is induction where the set of instances is not exhaustive. From a reasonable collection of instances, we sometimes leap to the conclusion that a schema is always true even though we have not seen all instances. Consider, for example, the function f where $f(1) = 1$, and $f(n + 1) = f(n) + 2n + 1$. If we look at some values of this function, we notice a certain regularity—the value of f always seems to be the square of its input. From this sample, we are tempted to leap to the conclusion that $f(n) = n^2$. Lucky guess. In this case, the conclusion happens to be true; and we can prove it.

n	$f(n)$	n^2
1	1	1
2	4	2^2
3	9	3^2
4	16	4^2
5	25	5^2

Here is another example. This one is due to the mathematician Fermat (1601–1665). He looked at values of the expression $2^{2^n} + 1$ for various values of n and noticed that they were all prime. So, he concluded the value of the expression was prime number. Unfortunately, this was not a lucky guess. His conjecture was ultimately disproved, in fact with the very next number in the sequence. (Mercifully, the counterexample was found after his death.)

n	$2^{2^n} + 1$	Prime?
1	5	Yes
2	17	Yes
3	257	Yes
4	65537	Yes

For us, this is not so good. In Logic, we are concerned with logical entailment. We want to derive only conclusions that are guaranteed to be true when the premises are true. Guesses like these are useful in suggesting possible conclusions, but they are not themselves proofs. In order to be sure of universally quantified conclusions, we must be sure that *all* instances are true. This is called *complete induction*. When applied to numbers, it is usually called *mathematical induction*.

The technique used for complete induction varies with the structure of the language to which it is applied. We begin this chapter with a discussion of domain closure, a rule that applies when the Herbrand base of our language is finite. We then move on to Linear Induction, i.e., the special case in which the ground terms in the language form a linear sequence. We look at tree induction, i.e., the special case in which the ground terms of the language form a tree. And we look at Structural Induction, which applies to all languages. Finally, we look at two special cases that make inductive reasoning more complicated—Multidimensional Induction and Embedded Induction.

11.2   DOMAIN CLOSURE

Induction for finite languages is trivial. We simply use the Domain Closure rule of inference. For a language with object constants $\sigma_1, \ldots, \sigma_n$, the rule is written as shown below. If we believe a schema is true for every instance, then we can infer a universally quantified version of that schema.

$$\varphi[\sigma_1]$$
$$\ldots$$
$$\frac{\varphi[\sigma_n]}{\forall v.\varphi[v]}$$

Recall that, in our formalization of the Sorority World, we have just four constants—*abby*, *bess*, *cody*, and *dana*. For this language, we would have the following Domain Closure rule.

Domain Closure (DC)
$$\varphi[abby]$$
$$\varphi[bess]$$
$$\varphi[cody]$$
$$\frac{\varphi[dana]}{\forall v.\varphi[v]}$$

The following proof shows how we can use this rule to derive an inductive conclusion. Given the premises we considered earlier in this book, it is possible to infer that Abby likes someone, Bess likes someone, Cody likes someone, and Dana likes someone. Taking these conclusions as premises and using our Domain Closure rule, we can then derive the inductive conclusion $\forall x.\exists y.likes(x,y)$, i.e., everybody likes somebody.

1.	$\exists y.likes(abby,y)$	Premise
2.	$\exists y.likes(bess,y)$	Premise
3.	$\exists y.likes(cody,y)$	Premise
4.	$\exists y.likes(dana,y)$	Premise
5.	$\forall x.\exists y.likes(x,y)$	Domain Closure: 1, 2, 3, 4

Unfortunately, this technique does not work when there are infinitely many ground terms. Suppose, for example, we have a language with ground terms σ_1, σ_2, ... A direct generalization of the Domain Closure rule is shown below.

$$\frac{\begin{array}{c}\varphi[\sigma_1] \\ \varphi[\sigma_2] \\ ...\end{array}}{\forall v.\varphi[v]}$$

This rule is sound in the sense that the conclusion of the rule is logically entailed by the premises of the rule. However, it does not help us produce a proof of this conclusion. To use the rule, we would need to prove all of the rule's premises. Unfortunately, there are infinitely many premises. So, the rule cannot be used in generating a finite proof.

All is not lost. It is sometimes possible to write rules that cover all instances without enumerating them individually. The method depends on the structure of the language. The next sections describe how this can be done for languages with different structures.

11.3 LINEAR INDUCTION

Imagine an infinite set of dominoes placed in a line so that, when one falls, the next domino in the line also falls. If the first domino falls, then the second domino falls. If the second domino falls, then the third domino falls. And so forth. By continuing this chain of reasoning, it is easy for us to convince ourselves that every domino eventually falls. This is the intuition behind a technique called Linear Induction.

Consider a language with a single object constant a and a single unary function constant s. There are infinitely many ground terms in this language, arranged in what resembles a straight line. See below. Starting from the object constant, we move to a term in which we apply the function constant to that constant, then to a term in which we apply the function constant to that term, and so forth.

$$a \rightarrow s(a) \rightarrow s(s(a)) \rightarrow s(s(s(a))) \rightarrow ...$$

In this section, we concentrate on languages that have linear structure of this sort. Hereafter, we call these *linear languages*. In all cases, there is a single object constant and a single unary function constant. In talking about a linear language, we call the object constant the *base element* of the language, and we call the unary function constant the *successor function*.

Although there are infinitely many ground terms in any linear language, we can still generate finite proofs that are guaranteed to be correct. The trick is to use the structure of the terms in the language in expressing the premises of an inductive rule of inference known as *Linear Induction*. See below for a statement of the induction rule for the language introduced above. In general, if we know that a schema holds of our base element and if we know that, whenever the schema holds of an element, it also holds of the successor of that element, then we can conclude that the schema holds of all elements.

Linear Induction

$$\varphi[a]$$
$$\frac{\forall \mu.(\varphi[\mu] \Rightarrow \varphi[s(\mu)])}{\forall \nu.\varphi[\nu]}$$

A bit of terminology before we go on. The first premise in this rule is called the *base case* of the induction, because it refers to the base element of the language. The second premise is called the *inductive case*. The antecedent of the inductive case is called the *inductive hypothesis*, and the consequent is called, not surprisingly, the *inductive conclusion*. The conclusion of the rule is sometimes called the *overall conclusion* to distinguish it from the inductive conclusion.

For the language introduced above, our rule of inference is sound. Suppose we know that a schema is true of a and suppose that we know that, whenever the schema is true of an arbitrary ground term τ, it is also true of the term $s(\tau)$. Then, the schema must be true of everything, since there are no other terms in the language.

The requirement that the signature consists of no other object constants or function constants is crucial. If this were not the case, say there were another object constant b, then we would have trouble. It would still be true that φ holds for every element in the set $\{a, s(a), s(s(a)),...\}$. However, because there are other elements in the Herbrand universe, e.g., b and $s(b)$, $\forall x.\varphi(x)$ would no longer be guaranteed.

Here is an example of induction in action. Recall the formalization of Arithmetic introduced in Chapter 9. Using the object constant 0 and the unary function constant s, we represent each number n by applying the function constant to 0 exactly n times. For the purposes of this example, let's assume we have just one ternary relation constant, viz. *plus*, which we use to represent the addition table.

The following axioms describe *plus* in terms of 0 and s. The first sentence here says that adding 0 to any element produces that element. The second sentence states that adding the successor of a number to another number yields the successor of their sum. The third sentence is a functionality axiom for *plus*.

$$\forall y.plus(0,y,y)$$
$$\forall x.\forall y.\forall z.(plus(x,y,z) \Rightarrow plus(s(x),y,s(z)))$$
$$\forall x.\forall y.\forall z.\forall w.(plus(x,y,z) \wedge \neg same(z,w) \Rightarrow \neg plus(x,y,w))$$

It is easy to see that any table that satisfies these axioms includes all of the usual addition facts. The first axiom ensures that all cases with 0 as first argument are included. From this fact and the second axiom, we can see that all cases with $s(0)$ as first argument are included. And so forth.

The first axiom above tells us that 0 is a *left identity* for addition—0 added to any number produces that number as result. As it turns out, given these definitions, 0 must also be a *right identity*, i.e., it must be the case that $\forall x.plus(x,0,x)$.

We can use induction to prove this result as shown below. We start with our premises. We use Universal Elimination on the first premise to derive the sentence on line 3. This takes care of the base case of our induction. We then start a subproof and assume the antecedent of the inductive case. We then use three applications of Universal Elimination on the second premise to get the sentence on line 5. We use Implication Elimination on this sentence and our assumption to derive the conclusion on line 6. We then discharge our assumption and form the implication shown on line 7 and then universalize this to get the result on line 8. Finally, we use Linear Induction to derive our overall conclusion.

1.	$\forall y.plus(0,y,y)$	Premise
2.	$\forall x.\forall y.\forall z.(plus(x,y,z) \Rightarrow plus(s(x),y,s(z)))$	Premise
3.	$plus(0,0,0)$	UE: 1
4.	$\quad plus(x,0,x)$	Assumption
5.	$\quad plus(x,0,x) \Rightarrow plus(s(x),0,s(x))$	$3 \times$ UE: 2
6.	$\quad plus(s(x),0,s(x))$	IE: 5, 4
7.	$plus(x,0,x) \Rightarrow plus(s(x),0,s(x))$	II: 4, 6
8.	$\forall x.(plus(x,0,x) \Rightarrow plus(s(x),0,s(x)))$	UI: 7
9.	$\forall x.(plus(x,0,x))$	Ind: 3, 8

Most inductive proofs have this simple structure. We prove the base case. We assume the inductive hypothesis; we prove the inductive conclusion; and, based on this proof, we have the inductive case. From the base case and the inductive case, we infer the overall conclusion.

11.4   TREE INDUCTION

Tree languages are a superset of linear languages. While in linear languages the terms in the language form a linear sequence, in tree languages the structure is more tree-like. Consider a language with an object constant a and two unary function constants f and g. Some of the terms in this language are shown below.

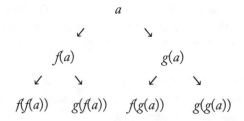

As with linear languages, we can write an inductive rule of inference for tree languages. The tree induction rule of inference for the language just described is shown below. Suppose we know that a schema φ holds of a. Suppose we know that, whenever the schema holds of any element, it holds of the term formed by applying f to that element. And suppose we know that, whenever the schema holds of any element, it holds of the term formed by applying g to that element. Then, we can conclude that the schema holds of every element.

Tree Induction

$$\varphi[a]$$
$$\forall\mu.(\varphi[\mu] \Rightarrow \varphi[f(\mu)])$$
$$\underline{\forall\mu.(\varphi[\mu] \Rightarrow \varphi[g(\mu)])}$$
$$\forall\nu.\varphi[\nu]$$

In order to see an example of tree induction in action, consider the ancestry tree for a particular dog. We use the object constant *rex* to refer to the dog; we use the unary function constant f to map an arbitrary dog to its father; and we use g map a dog to its mother. Finally, we have a single unary relation constant *purebred* that is true of a dog if and only if it is purebred.

Now, we write down the fundamental rule of dog breeding—we say that a dog is purebred if and only if both its father and its mother are purebred. See below. (This is a bit oversimplified on several grounds. Properly, the father and mother should be of the same breed. Also, this formalization suggests that every dog has an ancestry tree that stretches back without end. However, let's ignore these imperfections for the purposes of our example.)

$$\forall x.(purebred(x) \Leftrightarrow purebred(f(x)) \wedge purebred(g(x)))$$

Suppose now that we discover the fact that our dog *rex* is purebred. Then, we know that every dog in his ancestry tree must be purebred. We can prove this by a simple application of tree induction.

A proof of our conclusion is shown below. We start with the premise that Rex is purebred. We also have our constraint on purebred animals as a premise. We use Universal Elimination to instantiate the second premise, and then we use Biconditional Elimination on the biconditional in line 3 to produce the implication on line 4. On line 5, we start a subproof with the assumption the x is purebred. We use Implication Elimination to derive the conjunction on line 6, and then we use And Elimination to pick out the first conjunct. We then use Implication Introduction to discharge our assumption, and we Universal Introduction to produce the inductive case for f. We then repeat this process to produce an analogous result for g on line 14. Finally, we use the tree induction rule on the sentences on lines 1, 9, and 14 and thereby derive the desired overall conclusion.

1.	$purebred(rex)$	Premise
2.	$\forall x.(purebred(x) \Longleftrightarrow purebred(f(x)) \wedge purebred(g(x)))$	Premise
3.	$(purebred(x) \Longleftrightarrow purebred(f(x)) \wedge purebred(g(x)))$	UE: 2
4.	$(purebred(x) \Rightarrow purebred(f(x)) \wedge purebred(g(x)))$	BE: 3
5.	$\quad purebred(x)$	Assumption
6.	$\quad purebred(f(x)) \wedge purebred(g(x))$	IE: 4, 5
7.	$\quad purebred(f(x))$	AE: 6
8.	$purebred(x) \Rightarrow purebred(f(x))$	II: 5, 7
9.	$\forall x.purebred(x) \Rightarrow purebred(f(x))$	UI: 8
10.	$\quad purebred(x)$	Assumption
11.	$\quad purebred(f(x)) \wedge purebred(g(x))$	IE: 4, 10
12.	$\quad purebred(g(x))$	AE: 11
13.	$purebred(x) \Rightarrow purebred(g(x))$	II: 10, 12
14.	$\forall x.purebred(x) \Rightarrow purebred(g(x))$	UI: 13
15.	$\forall x.purebred(x)$	Ind: 1, 9, 14

11.5   STRUCTURAL INDUCTION

Structural Induction is the most general form of induction. In Structural Induction, we can have multiple object constants, multiple function constants, and, unlike our other forms of induction, we can have function constants with multiple arguments. Consider a language with two object constants a and b and a single binary function constant c. See below for a list of some of the terms in the language. We do not provide a graphical rendering in this case, as the structure is more complicated than a line or a tree.

$$a, b, c(a,a), c(a,b), c(b,a), c(b,b), c(a,c(a,a)), c(c(a,a),a), c(c(a,a),c(a,a)), ...$$

The Structural Induction rule for this language is shown below. If we know that φ holds of our base elements a and b and if we know $\forall \mu.\forall \nu.((\varphi[\mu] \wedge \varphi[\nu]) \Rightarrow \varphi[c(\mu,\nu)])$, then we can conclude $\forall \nu.\varphi[\nu]$ in a single step.

Structural Induction

$$\varphi[a]$$
$$\varphi[b]$$
$$\frac{\forall \lambda. \forall \mu.((\varphi[\lambda] \wedge \varphi[\mu]) \Rightarrow \varphi[c(\lambda,\mu)])}{\forall v.\varphi[v]}$$

As an example of a domain where Structural Induction is appropriate, recall the world of lists and trees introduced in Chapter 9. Let's assume we have two object constants a and b, a binary function constant c, and two unary relation constants p and q. Relation p is true of a structured object if and only if every fringe node is an a. Relation q is true of a structured object if and only if at least one fringe node is an a. The positive and negative axioms defining the relations are shown below.

$p(a)$ $q(a)$

$\forall u. \forall v.(p(u) \wedge p(v) \Rightarrow p(c(u,v)))$ $\forall u. \forall v.(q(u) \Rightarrow q(c(u,v)))$

$\neg p(b)$ $\forall u. \forall v.(q(v) \Rightarrow q(c(u,v)))$

$\forall u. \forall v.(p(c(u,v)) \Rightarrow p(u))$ $\neg q(b)$

$\forall u. \forall v.(p(c(u,v)) \Rightarrow p(v))$ $\forall u. \forall v.(q(c(u,v)) \Rightarrow q(u) \vee q(v))$

Now, as an example of Structural Induction in action, let's prove that every object that satisfies p also satisfies q. In other words, we want to prove the conclusion $\forall x.(p(x) \Rightarrow q(x))$. As usual, we start with our premises.

1. $p(a)$ Premise

2. $\forall u. \forall v.(p(u)) \wedge p(c(u,v)))$ Premise

3. $\neg p(b)$ Premise

4. $\forall u. \forall v.(p(c(u,v)) \Rightarrow p(u))$ Premise

5. $\forall u. \forall v.(p(c(u,v)) \Rightarrow p(v))$ Premise

6. $q(a)$ Premise

7. $\forall u. \forall v.(q(u) \Rightarrow q(c(u,v)))$ Premise

8. $\forall u. \forall v.(q(v) \Rightarrow q(c(u,v)))$ Premise

9. $\neg q(b)$ Premise

10. $\forall u. \forall v.(q(c(u,v)) \Rightarrow q(u) \vee q(v))$ Premise

To start the induction, we first prove the base cases for the conclusion. In this world, with two object constants, we need to show the result twice—once for each object constant in the language.

Let's start with a. The derivation is simple in this case. We assume $p(a)$, reiterate $q(a)$ from line 6, then use Implication Introduction to prove $(p(a) \Rightarrow q(a))$.

11.	$p(a)$	Assumption
12.	$q(a)$	Reiteration: 6
13.	$p(a) \Rightarrow q(a)$	Implication Introduction

Now, let's do the case for b. As before, we assume $p(b)$, and our goal is to derive $q(b)$. This is a little strange. We know that $q(b)$ is false. Still, we should be able to derive it since we have assumed $p(b)$, which is also false. The trick here is to generate contradictory conclusions from the assumption $\neg q(b)$. To this end, we assume $\neg q(b)$ and first prove $p(b)$. Having done so, we use Implication Introduction to get one implication. Then, we assume $\neg q(b)$ again and this time derive $\neg p(b)$ and get an implication. At this point, we can use Negation Introduction to derive $\neg\neg q(b)$ and Negation Elimination to get $q(b)$. Finally, we use Implication Introduction to prove $(p(b) \Rightarrow q(b))$.

14.	$p(b)$	Assumption
15.	$\neg q(b)$	Assumption
16.	$p(b)$	Reiteration: 14
17.	$\neg q(b) \Rightarrow p(b)$	Implication Introduction: 14, 16
18.	$\neg q(b)$	Assumption
19.	$\neg p(b)$	Reiteration: 3
20.	$\neg q(b) \Rightarrow \neg p(b)$	Implication Introduction: 18, 19
21.	$\neg\neg q(b)$	Negation Elimination: 17, 20
22.	$q(b)$	Negation Elimination: 21
23.	$p(b) \Rightarrow q(b)$	Implication Introduction: 17, 19

Having dealt with the base cases, the next step is to prove the inductive case. We need to show that, if our conclusion holds of u and v, then it also holds of $c(u,v)$. To this end, we assume the conjunction of our assumptions and then use And Elimination to split that conjunction into its two conjuncts. Our inductive conclusion is also an implication; and, to prove it, we assume its antecedent $p(c(u,v))$. From this, we derive $p(u)$; from that, we derive $q(u)$; and, from that, we derive $q(c(u,v))$. We then use Implication Introduction to get the desired implication. A final use of Implication Introduction and a couple of applications of Universal Introduction gives us the inductive case for the induction.

24.	$(p(u) \Rightarrow q(u)) \wedge (p(v) \Rightarrow q(v))$	Assumption
25.	$p(u) \Rightarrow q(u)$	And Elimination: 24
26.	$p(v) \Rightarrow q(v)$	And Elimination: 24
27.	$p(c(u,v))$	Assumption
28.	$\forall v.(p(c(u,v)) \Rightarrow p(u))$	Universal Elimination: 4
29.	$p(c(u,v)) \Rightarrow p(u)$	Universal Elimination: 28
30.	$p(u)$	Implication Elimination: 29, 27
31.	$q(u)$	Implication Elimination: 25, 30
32.	$\forall v.(q(u) \Rightarrow q(c(u,v)))$	Universal Elimination: 7
33.	$q(u) \Rightarrow q(c(u,v))$	Universal Elimination: 32
34.	$q(c(u,v))$	Implication Elimination: 33, 31
35.	$p(c(u,v)) \Rightarrow q(c(u,v))$	Implication Elimination: 27, 34
36.	$(p(u) \Rightarrow q(u)) \wedge (p(v) \Rightarrow q(v)) \Rightarrow p(c(u,v)) \Rightarrow q(c(u,v))$	Implication Elimination: 24, 35
37.	$\forall v.((p(u) \Rightarrow q(u)) \wedge (p(v) \Rightarrow q(v)) \Rightarrow p(c(u,v))) \Rightarrow q(c(u,v)))$	Universal Introduction: 36
38.	$\forall u.\,\forall v.((p(u) \Rightarrow q(u)) \wedge (p(v) \Rightarrow q(v)) \Rightarrow p(c(u,v))) \Rightarrow q(c(u,v)))$	Universal Introduction: 37

Finally, using the base case on lines 13 and 23 and the inductive case on line 38, we use Structural Induction to give us the conclusion we set out to prove.

39.	$\forall x.(p(x) \Rightarrow q(x))$	Induction: 13, 23, 38

Although the proof in this case is longer than in the previous examples, the basic inductive structure is the same. Importantly, using induction, we can prove this result where otherwise it would not be possible.

11.6   MULTIDIMENSIONAL INDUCTION

In our look at induction thus far, we have been concentrating on examples in which the conclusion is a universally quantified sentence with just one variable. In many situations, we want to use induction to prove a result with more than one universally quantified variable. This is called *multidimensional induction* or, sometimes, *multivariate induction*.

In principle, multidimensional induction is straightforward. We simply use ordinary induction to prove the outermost universally quantified sentence. Of course, in the case of multidimensional induction the base case and the inductive conclusion are themselves universally quantified sentences; and, if necessary we use induction to prove these subsidiary results.

As an example, consider a language with a single object constant a, a unary function constant s, and a binary relation constant e. The axioms shown below define e.

$$e(a,a)$$
$$\forall x.\neg e(a,s(x))$$
$$\forall x.\neg e(s(x),a)$$
$$\forall x.\forall y.(e(x,y) \Rightarrow e(s(x),s(y)))$$
$$\forall x.\forall y.(e(s(x),s(y)) \Rightarrow e(x,y))$$

The relation e is an equivalence relation—it is reflexive, symmetric, and transitive. Proving reflexivity is easy. Proving transitivity is left as an exercise for the reader. Our goal here is to prove symmetry.

Goal: $\forall x.\forall y.(e(x,y) \Rightarrow e(y,x))$.

In what follows, we use induction to prove the outer quantified formula and then use induction on each of the inner conclusions as well. This means we have two immediate subgoals—the base case for the outer induction and the inductive case for the outer induction.

Goal: $\forall y.(e(a,y) \Rightarrow e(y,a))$.
Goal: $\forall x.(\forall y.(e(x,y) => e(y,x)) \Rightarrow \forall y.(e(s(x),y) => e(y,s(x))))$.

As usual, we start with our premises. We then prove the base case of the inner induction. This is easy. We assume $e(a,a)$ and then use Implication Introduction to prove the base case in one step.

1.	$e(a, a)$	Premise
2.	$\forall x.\neg e(a,s(x))$	Premise
3.	$\forall x.\neg e(s(x),a)$	Premise
4.	$\forall x.\forall y.(e(x,y) \Rightarrow (e(s(x), s(y))))$	Premise
5.	$\forall x.\forall y.(e(s(x), s(y)) \Rightarrow e(x, y))$	Premise
6.	$\quad\mid e(a, a)$	Assumption
7.	$e(a, a) \Rightarrow e(a, a)$	Implication Introduction: 6, 6

The next step is to prove the inductive case for the inner induction. To this end, we assume the inductive hypothesis and try to prove the inductive conclusion. Since the conclusion is itself an implication, we assume its antecedent and prove the consequent. As shown below, we do this by assuming the consequent is false and proving a sentence and its negation. We then use Negation Introduction and Negation Elimination to derives the consequent. We finish with two applications of Implication Introduction and an application of Universal Introduction.

8.	$e(a, y) \Rightarrow e(y, a))$	Assumption
9.	$e(a,s(y))$	Assumption
10.	$\neg e(s(y),a)$	Assumption
11.	$e(a,s(y))$	Reiteration: 9
12.	$\neg e(s(y),a) \Rightarrow e(a,s(y))$	Implication Introduction: 10, 11
13.	$\neg e(s(y),a)$	Assumption
14.	$\neg e(a,s(y))$	Universal Instantiation: 2
15.	$\neg e(s(y),a) \Rightarrow \neg e(a,s(y))$	Implication Introduction: 13, 14
16.	$\neg\neg e(s(y),a)$	Negation Elimination: 12, 15
17.	$e(s(y),a)$	Negation Elimination: 16
18.	$e(a, s(y)) \Rightarrow e(s(y),a)$	Implication Introduction: 9, 17
19.	$e(a, y) \Rightarrow e(y, a)) \Rightarrow (e(a,s(y)) \Rightarrow e(s(y),a))$	Implication Introduction: 8, 18
20.	$\forall y.((e(a, y) \Rightarrow e(y, a)) \Rightarrow (e(a,s(y)) \Rightarrow e(s(y),a)))$	Universal Introduction: 19
21.	$\forall y.(e(a, y) \Rightarrow e(y, a))$	Induction: 7, 20

That's a lot of work just to prove the base case of the outer induction. The inductive case of the outer induction is even more complex, and it is easy to make mistakes. The trick to avoiding these mistakes is to be methodical.

In order to prove the inductive case for the outer induction, we assume the inductive hypothesis $\forall y.(e(x,y) \Rightarrow e(y,x))$; and we then prove the inductive conclusion $\forall y.(e(s(x),y) \Rightarrow e(y,s(x))))$. We prove this by induction on the variable y.

We start by proving the base case for this inner induction. We start with the inductive hypothesis. We then assume the antecedent of the base case.

22.	$\forall y.\ e(x, y) \Rightarrow e(y, x))$	Assumption
23.	$e(s(yx),a)$	Assumption
24.	$\neg e(a,s(x))$	Assumption
25.	$e(s(x),a)$	Reiteration: 23
26.	$\neg e(a,s(x))) \Rightarrow e(s(x),a)$	Implication Introduction: 24, 25
27.	$\neg e(a,s(x))$	Assumption
28.	$\neg e(s(x),a)$	Universal Elimination: 3
29.	$\neg e(a,s(x))) \Rightarrow \neg e(s(x),a)$	Implication Introduction: 27, 28
30.	$\neg\neg e(a,s(x)))$	Negation Elimination: 26, 29
31.	$e(a,s(x)))$	Negation Elimination: 30
32.	$e(s(x),a) \qquad e(a,s(x)))$	Implication Introduction: 23, 31

Next, we work on the inductive case for the second inner induction. We start by assuming the inductive hypothesis. We then assume the antecedent of the inductive conclusion.

33.	$e(s(x),y) \Rightarrow e(y,s(x))$	Assumption
34.	$e(s(x),s(y))$	Assumption
35.	$\forall y.(e(s(x)),s(y)) \Rightarrow e(x,y))$	Universal Elimination: 5
36.	$e(s(x),s(y))) \Rightarrow e(x,y)$	Universal Elimination: 35
37.	$e(x,y)$	Implication Elimination: 36, 34
38.	$e(x,y) \Rightarrow e(y,x)$	Universal Elimination: 22
39.	$e(y,x)$	Implication Elimination: 38, 37
40.	$\forall y.(e(y,x) \Rightarrow e(s(y),s(x)))$	Universal Elimination: 4
41.	$e(y,x) \Rightarrow e(s(y),s(x))$	Universal Elimination: 40
42.	$e(s(y),s(x))$	Implication Elimination: 41, 39
43.	$e(s(x)s(y)) \Rightarrow e(s,(y),s(x))$	Implication Introduction: 34, 42
44.	$e(s(x),y) \Rightarrow e(y,s(x))) \Rightarrow (e(s(x),s(y)) \Rightarrow e(s,(y),s(x)))$	Implication Introduction: 33, 43
45.	$\forall y.((e(s(x),y) \Rightarrow e(y,s(x))) \Rightarrow (e(s(x),s(y)) \Rightarrow e(s,(y),s(x))))$	Universal Introduction: 43

From the results on lines 32 and 45, we can conclude the inductive case for the outer induction.

46.	$\forall y.(e(s(x),y) \Rightarrow e(y,s(x)))$	Induction: 32, 45
47.	$\forall y.(e(x,y) \Rightarrow e(y,x)) \Rightarrow \forall y.(e(s(x),y) \Rightarrow e(y,s(x)))$	Implication Introduction: 22, 46
48.	$\forall x.(\forall y.(e(x,y) \Rightarrow e(y,x)) \Rightarrow \forall y.(e(s(x),y) \Rightarrow e(y,s(x)))$	Universal Introduction: 47

Finally, from the base case for the outer induction and this inductive case, we can conclude our overall result.

49. $\forall x.\forall y.(e(x,y) \Rightarrow e(y,x))$ Induction: 7, 48

As this proof illustrates, the technique of using induction within induction works just fine. Unfortunately, it is tedious and error-prone. for this reason, many people prefer to use specialized forms of multidimensional induction.

11.7   EMBEDDED INDUCTION

In all of the examples in this chapter thus far, induction is used to prove the overall result. While this approach works nicely in many cases, it is not always successful. In some cases, it is easier to use induction on parts of a problem or to prove alternative conclusions and then use these intermediate results to derive the overall conclusions (using inductive or non-inductive methods).

As an example, consider a world characterized by a single object constant a, a single unary function constant s, and a single unary relation constant p. Assume we have the set of axioms shown below.

$$\forall x.(p(x) \Rightarrow p(s(s(x))))$$
$$p(a)$$
$$p(s(a))$$

A little thought reveals that these axioms logically entail the universal conclusion $\forall x.p(x)$. Unfortunately, we cannot derive this conclusion directly using Linear Induction. The base case is easy enough. And, from $p(x)$ we can easily derive $p(s(s(x)))$. However, it is not so easy to derive $p(s(x))$, which is what we need for the inductive case of Linear Induction.

The good news is that we can succeed in cases like this by proving a slightly more complicated intermediate conclusion and then using that conclusion to prove the result. One way to do this is shown below. In this case, we start by using Linear Induction to prove $\forall x.(p(x) \wedge p(s(x)))$. The base case $p(a) \wedge p(s(a))$ is easy, since we are given the two conjuncts as axioms. The inductive case is straightforward. We assume $p(x) \wedge p(s(x))$. From this hypothesis, we use And Elimination to get $p(x)$ and $p(s(x))$. We then use Universal Elimination and Implication Elimination to derive $p(s(s(x)))$. We then conjoin these results, use Implication Introduction and Universal Introduction to get the inductive case for our induction. From the base case and the inductive case, we get our intermediate conclusion. Finally, starting with this conclusion, we use Universal Elimination, And Elimination, and Universal Introduction to get the overall result.

1.	$\forall y.(p(x) \Rightarrow p(s(s(x))))$	Premise
2.	$p(a)$	Premise
3.	$p(s(a))$	Premise
4.	$p(a) \wedge p(s(a))$	And Introduction
5.	$\quad\mid p(x) \wedge p(s(x))$	Assumption
6.	$\quad\mid p(x)$	And Elimination
7.	$\quad\mid p(s(x))$	And Elimination
8.	$\quad\mid p(x) \Rightarrow p(s(s(x)))$	Universal Elimination: 1
9.	$\quad\mid p(s(s(x)))$	Implication Elimination: 8, 6
10.	$\quad\mid p(s(x)) \wedge p(s(s(x)))$	And Introduction: 7, 9
11.	$p(x) \wedge p(s(x)) \Rightarrow p(s(x)) \wedge p(s(s(x)))$	Implication Introduction: 5, 10
12.	$\forall x.(p(x) \wedge p(s(x))) \Rightarrow p(s(x)) \wedge p(s(s(x)))$	Universal Introduction: 11
13.	$\forall x.(p(x) \wedge p(s(x)))$	Induction: 4, 12
14.	$p(x) \wedge p(s(x))$	Universal Elimination: 13
15.	$p(x)$	And Elimination: 14
16.	$p(s(x))$	And Elimination: 14
17.	$\forall x.p(x)$	Universal Introduction: 16

In this case, we are lucky that there is a useful conclusion that we can prove with standard Linear Induction. Things are not always so simple; and in some cases we need more complex forms of induction. Unfortunately, there is no finite collection of approaches to induction that covers all cases. If there were, we could build an algorithm for determining logical entailment for Herbrand Logic in all cases; and, as we discussed in Chapter 10, there is no such algorithm.

RECAP

Induction is reasoning from the specific to the general. *Complete induction* is induction where the set of instances is exhaustive. *Incomplete induction* is induction where the set of instances is not exhaustive. *Linear Induction* is a type of complete induction for languages with a single object constants and a single unary function constant. *Tree Induction* is a type of complete induction for languages with a single object constants and multiple unary function constants. *Structural Induction* is a generalization of both Linear Induction and Tree Induction that works even in the presence of multiple object constants and multiple *n*-ary function constants.

11.8 EXERCISES

11.1. Assume a language with the object constants *a* and *b* and no function constants. Given $q(a)$ and $q(b)$, use the Fitch system with domain closure to prove $\forall x.(p(x) \Rightarrow q(x))$.

11.2. Assume a language with the object constant a and the function constant s. Given $r(a)$, $\forall x.(p(x) \Rightarrow r(s(x)))$, $\forall x.(q(x) \Rightarrow r(s(x)))$, and $\forall x.(r(x) \Rightarrow p(x) \vee q(x))$, use the Fitch system with Linear Induction to prove $\forall x.r(x)$.

11.3. Assume a language with object constant a and unary function constants f and g. Given $p(a)$, $\forall x.(p(x) \Rightarrow p(f(x)))$, and $\forall x.(p(f(x)) \Rightarrow p(g(x)))$, use the Fitch system with Tree Induction to prove $\forall x.p(x)$.

11.4. Consider a language with object constants a and b, binary function constant c, and unary relation constants m and p and q. The definitions for the relations are shown below. Relation m is true of a and only a. Relation p is true of a structured object if and only if it is a linear list (as defined in Chapter 9) with a top-level element that satisfies m. Relation q is true of a structured object if and only if there is an element anywhere in the structure that satisfies m.

$m(a)$	$\forall u.\forall v.(m(u) \Rightarrow p(c(u,v)))$	$\forall u.(m(u) \Rightarrow q(u)))$
$\neg m(b)$	$\forall u.\forall v.(p(v) \Rightarrow p(c(u,v)))$	$\forall u.\forall v.(q(u)$
		$\Rightarrow q(c(u,v)))$
$\forall u.\forall v.\neg m(c(u,v))$	$\neg p(a)$	$\forall u.\forall v.(q(v)$
		$\Rightarrow q(c(u,v)))$
	$\neg p(b)$	$\neg m(a) \Rightarrow \neg q(a)$
	$\forall u.\forall v.(p(c(u,v)) \Rightarrow m(u) \vee p(v))$	$\neg m(b) \Rightarrow \neg q(b)$
		$\forall u.\forall v.(q(c(u,v))$
		$\Rightarrow q(u) \vee q(v))$

Your job is to show that any object that satisfies p also satisfies q. Starting with the preceding axioms, use Fitch with Structural Induction to prove $\forall x.(p(x) \Rightarrow q(x))$. Beware: The proof requires more than 50 steps (including the premises). The good news is that it is very similar to the proof in Section 11.5.

11.5. Starting with the axioms for e given in Section 11.6, it is possible to prove that e is transitive, i.e., $\forall x.\forall y.\forall z.(e(x,y) \wedge e(y,z) \Rightarrow e(x,z))$. Doing this requires a three variable induction, and it is quite messy. Your job in this problem is to prove just the base case for the outermost induction, i.e., prove $\forall y.\forall z.(e(a,y) \wedge e(y,z) \Rightarrow e(a,z))$. Hint: Use the strategy illustrated in Section 11.6. Extra credit: Do the full proof of transitivity.

11.6. Consider a language with a single object constant a, a single unary function constant s, and two unary relation constants p and q. We start with the premises shown below. We know that p is true of $s(a)$ and only $s(a)$. We know that q is also true of $s(a)$, but we do not know whether it is true of anything else.

$\neg p(a)$
$p(s(a))$
$\forall x.\neg p(s(s(x)))$
$q(s(a))$

Prove $\forall x.(p(x) \Rightarrow q(x))$. Hint: Break the problem into two parts—first prove the result for $s(x)$, and then use that intermediate conclusion to prove the overall result.

CHAPTER 12

Resolution

12.1 INTRODUCTION

The *Resolution Principle* is a rule of inference for Relational Logic analogous to the Propositional Resolution Principle for Propositional Logic. Using the Resolution Principle alone (without axiom schemata or other rules of inference), it is possible to build a reasoning program that is sound and complete for all of Relational Logic. The search space using the Resolution Principle is smaller than the search space for generating Herbrand proofs.

In our tour of resolution, we look first at unification, which allows us to *unify* expressions by substituting terms for variables. We then move on to a definition of clausal form extended to handle variables. The Resolution Principle follows. We then look at some applications. Finally, we examine strategies for making the procedure more efficient.

12.2 CLAUSAL FORM

As with Propositional Resolution, Resolution works only on expressions in *clausal form*. The definitions here are analogous. A *literal* is either a relational sentence or a negation of a relational sentence. A *clause* is a set of literals and, as in Propositional Logic, represents a disjunction of the literals in the set. A clause set is a set of clauses and represents a conjunction of the clauses in the set.

The procedure for converting relational sentences to clausal form is similar to that for Propositional Logic. Some of the rules are the same. However, there are a few additional rules to deal with the presence of variables and quantifiers. The conversion rules are summarized below and should be applied in order.

In the first step (Implications out), we eliminate all occurrences of the $\Rightarrow$, $\Leftarrow$, and $\Leftrightarrow$ operators by substituting equivalent sentences involving only the $\wedge$, $\vee$, and $\neg$ operators.

$$\begin{array}{rcl} \varphi \Rightarrow \psi & \rightarrow & \neg\varphi \vee \psi \\ \varphi \Leftarrow \psi & \rightarrow & \varphi \vee \neg\psi \\ \varphi \Leftrightarrow \psi & \rightarrow & (\neg\varphi \vee \psi) \wedge (\varphi \vee \neg\psi) \end{array}$$

In the second step (Negations in), negations are distributed over other logical operators and quantifiers until each such operator applies to a single atomic sentence. The following replacement rules do the job.

$$\neg\neg\varphi \quad\quad\quad\quad \rightarrow \quad \varphi$$
$$\neg(\varphi \wedge \psi) \quad \rightarrow \quad \neg\varphi \vee \neg\psi$$
$$\neg(\varphi \vee \psi) \quad \rightarrow \quad \neg\varphi \wedge \neg\psi$$
$$\neg\forall v.\varphi \quad\quad \rightarrow \quad \exists v.\neg\varphi$$
$$\neg\exists v.\varphi \quad\quad \rightarrow \quad \forall v.\neg\varphi$$

In the third step (Standardize variables), we rename variables so that each quantifier has a unique variable, i.e., the same variable is not quantified more than once within the same sentence. The following transformation is an example.

$$\forall x.(p(x) \Rightarrow \exists x.q(x)) \quad \rightarrow \quad \forall x.(p(x) \Rightarrow \exists y.q(y))$$

In the fourth step (Existentials out), we eliminate all existential quantifiers. The method for doing this is a little complicated, and we describe it in two stages.

If an existential quantifier does not occur within the scope of a universal quantifier, we simply drop the quantifier and replace all occurrences of the quantified variable by a new constant; i.e., one that does not occur anywhere else in our database. The constant used to replace the existential variable in this case is called a *Skolem constant*. The following example assumes that a is not used anywhere else.

$$\exists x.p(x) \quad \rightarrow \quad p(a)$$

If an existential quantifier is within the scope of any universal quantifiers, there is the possibility that the value of the existential variable depends on the values of the associated universal variables. Consequently, we cannot replace the existential variable with a constant. Instead, the general rule is to drop the existential quantifier and to replace the associated variable by a term formed from a new function symbol applied to the variables associated with the enclosing universal quantifiers. Any function defined in this way is called a *Skolem function*. The following example illustrates this transformation. It assumes that f is not used anywhere else.

$$\forall x.(p(x) \wedge \exists z.q(x, y, z)) \quad \rightarrow \quad \forall x.(p(x) \wedge q(x, y, f(x, y)))$$

In the fifth step (Alls out), we drop all universal quantifiers. Because the remaining variables at this point are universally quantified, this does not introduce any ambiguities.

$$\forall x.(p(x) \wedge q(x, y, f(x, y))) \quad \rightarrow \quad p(x) \wedge q(x, y, f(x, y))$$

In the sixth step (Disjunctions in), we put the expression into *conjunctive normal form*, i.e., a conjunction of disjunctions of literals. This can be accomplished by repeated use of the following rules.

$$\varphi \vee (\psi \wedge \chi) \quad\quad\quad \rightarrow \quad (\varphi \vee \psi) \wedge (\varphi \vee \chi)$$
$$(\varphi \wedge \psi) \vee \chi \quad\quad\quad \rightarrow \quad (\varphi \vee \chi) \wedge (\psi \vee \chi)$$
$$\varphi \vee (\varphi_1 \vee ... \vee \varphi_n) \quad \rightarrow \quad \varphi \vee \varphi_1 \vee ... \vee \varphi_n$$
$$(\varphi_1 \vee ... \vee \varphi_n) \vee \varphi \quad \rightarrow \quad \varphi_1 \vee ... \vee \varphi_n \vee \varphi$$
$$\varphi \wedge (\varphi_1 \wedge ... \wedge \varphi_n) \quad \rightarrow \quad \varphi \wedge \varphi_1 \wedge ... \wedge \varphi_n$$
$$(\varphi_1 \wedge ... \wedge \varphi_n) \wedge \varphi \quad \rightarrow \quad \varphi_1 \wedge ... \wedge \varphi_n \wedge \varphi$$

In the seventh step (Operators out), we eliminate operators by separating any conjunctions into its conjuncts and writing each disjunction as a separate clause.

$$\varphi_1 \wedge ... \wedge \varphi_n \quad \rightarrow \quad \varphi_1$$
$$\rightarrow \quad ...$$
$$\rightarrow \quad \varphi_n$$
$$\varphi_1 \vee ... \vee \varphi_n \quad \rightarrow \quad \{\varphi_1, ... , \varphi_n\}$$

As an example of this conversion process, consider the problem of transforming the following expression to clausal form. The initial expression appears on the top line, and the expressions on the labeled lines are the results of the corresponding steps of the conversion procedure.

$$\exists y.(g(y) \wedge \forall z.(r(z) \Rightarrow f(y, z)))$$

I $\exists y.(g(y) \wedge \forall z.(\neg r(z) \vee f(y, z)))$

N $\exists y.(g(y) \wedge \forall z.(\neg r(z) \vee f(y, z)))$

S $\exists y.(g(y) \wedge \forall z.(\neg r(z) \vee f(y, z)))$

E $g(gary) \wedge \forall z.(\neg r(z) \vee f(gary, z))$

A $g(gary) \wedge (\neg r(z) \vee f(gary, z))$

D $g(gary) \wedge (\neg r(z) \vee f(gary, z))$

O $\{g(gary)\}$
     $\{\neg r(z), f(gary, z)\}$

Here is another example. In this case, the starting sentence is almost the same. The only difference is the leading $\neg$, but the result looks quite different.

$$\neg \exists y.(g(y) \wedge \forall z.(r(z) \Rightarrow f(y, z)))$$

I $\neg \exists y.(g(y) \wedge \forall z.(\neg r(z) \vee f(y, z)))$

N $\forall y.(\neg(g(y) \wedge \forall z.(\neg r(z) \vee f(y, z)))$
     $\forall y.(\neg g(y) \vee \neg \forall z.(\neg r(z) \vee f(y, z)))$
     $\forall y.(\neg g(y) \vee \exists z.\neg(\neg r(z) \vee f(y, z)))$
     $\forall y.(\neg g(y) \vee \exists z.(\neg\neg r(z) \wedge \neg f(y, z)))$
     $\forall y.(\neg g(y) \vee \exists z.(r(z) \wedge \neg f(y, z)))$

S $\forall y.(\neg g(y) \vee \exists z.(r(z) \wedge \neg f(y, z)))$

E $\forall y.(\neg g(y) \vee (r(k(y)) \wedge \neg f(y, k(y))))$

A $\neg g(y) \vee (r(k(y)) \wedge \neg f(y, k(y)))$

D $(\neg g(y) \vee r(k(y))) \wedge (\neg g(y) \vee \neg f(y, k(y)))$

O $\{\neg g(y) \vee r(k(y))\}$
     $\{\neg g(y) \vee \neg f(y, k(y))\}$

In Propositional Logic, the clause set corresponding to any sentence is logically equivalent to that sentence. In Relational Logic, this is not necessarily the case. For example, the clausal form of the sentence $\exists x.p(x)$ is $\{p(a)\}$. This is not logically equivalent. It is not even in the same language. Since the clause exists in a language with an additional object constant, there are truth assignments that satisfy the sentence but not the clause. On the other hand, the converted clause set has a special relationship to the original set of sentences: over the expanded language, the clause set is satisfiable if and only if the original sentence is satisfiable (also over the expanded language). As we shall see, in resolution, this equivalence of satisfiability is all we need to obtain a proof method as powerful as the Fitch system presented in Chapter 7.

12.3   UNIFICATION

What differentiates Resolution from propositional resolution is unification. In propositional resolution, two clauses resolve if they contain complementary literals, i.e., the positive literal is identical to the target of the negative literal. The same idea underlies Resolution, except that the criterion for complementarity is relaxed. The positive literal does not need to be identical to the target of the negative literal; it is sufficient that the two can be made identical by substitutions for their variables.

Unification is the process of determining whether two expressions can be *unified*, i.e., made identical by appropriate substitutions for their variables. As we shall see, making this determination is an essential part of resolution.

A *substitution* is a finite mapping of variables to terms. In what follows, we write substitutions as sets of replacement rules, like the one shown below. In each rule, the variable to which the arrow is pointing is to be replaced by the term from which the arrow is pointing. In this case, x is to be replaced by a, y is to be replaced by $f(b)$, and z is to be replaced by v.

$$\{x \leftarrow a, y \leftarrow f(b), z \leftarrow v\}$$

The variables being replaced together constitute the *domain* of the substitution, and the terms replacing them constitute the *range*. For example, in the preceding substitution, the domain is $\{x, y, z\}$, and the range is $\{a, f(b), v\}$.

A substitution is *pure* if and only if all replacement terms in the range are free of the variables in the domain of the substitution. Otherwise, the substitution is *impure*. The substitution shown above is pure whereas the one shown below is impure.

$$\{x \leftarrow a, y \leftarrow f(b), z \leftarrow x\}$$

The result of applying a substitution σ to an expression φ is the expression $\varphi\sigma$ obtained from the original expression by replacing every occurrence of every variable in the domain of the substitution by the term with which it is associated.

$$
\begin{aligned}
q(x, y)\{x \leftarrow a, y \leftarrow f(b), z \leftarrow v\} &= q(a, f(b)) \\
q(x, x)\{x \leftarrow a, y \leftarrow f(b), z \leftarrow v\} &= q(a, a) \\
q(x, w)\{x \leftarrow a, y \leftarrow f(b), z \leftarrow v\} &= q(a, w) \\
q(z, v)\{x \leftarrow a, y \leftarrow f(b), z \leftarrow v\} &= q(v, v)
\end{aligned}
$$

Note that, if a substitution is pure, application is idempotent, i.e., applying a substitution a second time has no effect.

$$q(x, x, y, w, z)\{x \leftarrow a, y \leftarrow f(b), z \leftarrow v\} = q(a, a, f(b), w, v)$$
$$q(a, a, f(b), w, v)\{x \leftarrow a, y \leftarrow f(b), z \leftarrow v\} = q(a, a, f(b), w, v)$$

However, this is not the case for impure substitutions, as illustrated by the following example. Applying the substitution once leads to an expression with an x, allowing for a different answer when the substitution is applied a second time.

$$q(x, x, y, w, z)\{x \leftarrow a, y \leftarrow f(b), z \leftarrow x\} = q(a, a, f(b), w, x)$$
$$q(a, a, f(b), w, x)\{x \leftarrow a, y \leftarrow f(b), z \leftarrow x\} = q(a, a, f(b), w, a)$$

Given two or more substitutions, it is possible to define a single substitution that has the same effect as applying those substitutions in sequence. For example, the substitutions $\{x \leftarrow a, y \leftarrow f(u), z \leftarrow v\}$ and $\{u \leftarrow d, v \leftarrow e\}$ can be combined to form the single substitution $\{x \leftarrow a, y \leftarrow f(d), z \leftarrow e, u \leftarrow d, v \leftarrow e\}$, which has the same effect as the first two substitutions when applied to any expression whatsoever.

Computing the *composition* of a substitution σ and a substitution τ is easy. There are two steps. (1) First, we apply τ to the range of σ. (2) Then we adjoin to σ all pairs from τ with different domain variables.

As an example, consider the composition shown below. In the right hand side of the first equation, we have applied the second substitution to the replacements in the first substitution. In the second equation, we have combined the rules from this new substitution with the non-conflicting rules from the second substitution.

$$\{x \leftarrow a, y \leftarrow f(u), z \leftarrow v\}\{u \leftarrow d, v \leftarrow e, z \leftarrow g\}$$
$$= \{x \leftarrow a, y \leftarrow f(d), z \leftarrow e\}\{u \leftarrow d, v \leftarrow e, z \leftarrow g\}$$
$$= \{x \leftarrow a, y \leftarrow f(d), z \leftarrow e, u \leftarrow d, v \leftarrow e\}$$

It is noteworthy that composition does not necessarily preserve substitutional purity. The composition of two impure substitutions may be pure, and the composition of two pure substitutions may be impure.

This problem does not occur if the substitutions are *composable*. A substitution σ and a substitution τ are *composable* if and only if the domain of σ and the range of τ are disjoint. Otherwise, they are *noncomposable*.

$$\{x \leftarrow a, y \leftarrow b, z \leftarrow v\}\{x \leftarrow u, v \leftarrow b\}$$

By contrast, the following substitutions are noncomposable. Here, x occurs in both the domain of the first substitution and the range of the second substitution, violating the definition of composability.

$$\{x \leftarrow a, y \leftarrow b, z \leftarrow v\}\{x \leftarrow u, v \leftarrow x\}$$

The importance of composability is that it ensures preservation of purity. The composition of composable pure substitutions must be pure. In the sequel, we look only at compositions of composable pure substitutions.

A substitution σ is a *unifier* for an expression φ and an expression ψ if and only if $\varphi\sigma=\psi\sigma$, i.e., the result of applying σ to φ is the same as the result of applying σ to ψ. If two expressions have a unifier, they are said to be *unifiable*. Otherwise, they are *nonunifiable*.

The expressions $p(x, y)$ and $p(a,v)$ have a unifier, e.g., $\{x \leftarrow a, y \leftarrow b, v \leftarrow b\}$ and are, therefore, unifiable. The results of applying this substitution to the two expressions are shown below.

$$p(x, y)\{x \leftarrow a, y \leftarrow b, v \leftarrow b\}=p(a, b)$$
$$p(a, v)\{x \leftarrow a, y \leftarrow b, v \leftarrow b\}=p(a, b)$$

Note that, although this substitution unifies the two expressions, it is not the only unifier. We do not have to substitute b for y and v to unify the two expressions. We can equally well substitute c or d or $f(c)$ or $f(w)$. In fact, we can unify the expressions without changing v at all by simply replacing y by v.

In considering these alternatives, it should be clear that some substitutions are more general than others. We say that a substitution σ is *as general as or more general than* a substitution τ if and only if there is another substitution δ such that $\sigma\delta = \tau$. For example, the substitution $\{x \leftarrow a, y \leftarrow v\}$ is more general than $\{x \leftarrow a, y \leftarrow f(c), v \leftarrow f(c)\}$ since there is a substitution $\{v \leftarrow f(c)\}$ that, when applied to the former, gives the latter.

$$\{x \leftarrow a, y \leftarrow v\}\{v \leftarrow f(c)\}=\{x \leftarrow a, y \leftarrow f(c), v \leftarrow f(c)\}$$

In resolution, we are interested only in unifiers with maximum generality. A *most general unifier*, or *mgu*, σ of two expressions has the property that it as general as or more general than any other unifier.

Although it is possible for two expressions to have more than one most general unifier, all of these most general unifiers are structurally the same, i.e., they are unique up to variable renaming. For example, $p(x)$ and $p(y)$ can be unified by either the substitution $\{x \leftarrow y\}$ or the substitution $\{y \leftarrow x\}$; and either of these substitutions can be obtained from the other by applying a third substitution. This is not true of the unifiers mentioned earlier.

One good thing about our language is that there is a simple and inexpensive procedure for computing a most general unifier of any two expressions if it exists.

The procedure assumes a representation of expressions as sequences of subexpressions. For example, the expression $p(a, f(b),z)$ can be thought of as a sequence with four elements, viz. the relation constant p, the object constant a, the term $f(b)$, and the variable z. The term $f(b)$ can in turn be thought of as a sequence of two elements, viz. the function constant f and the object constant b.

We start the procedure with two expressions and a substitution, which is initially the empty substitution. We then recursively process the two expressions, comparing the subexpressions at each point. Along the way, we expand the substitution with variable assignments as described below. If, we fail to unify any pair of subexpression at any point in this process, the procedure as a whole fails. If we finish this recursive comparison of the expressions, the procedure as a whole succeeds, and the accumulated substitution at that point is the most general unifier.

In comparing two subexpressions, we first apply the substitution to each of the two expressions; and we then execute the following procedure on the two modified expressions.

1. If two modified expressions being compared are identical, then nothing more needs to be done.

2. If two modified expressions are not identical and both expressions are constants, then we fail, since there is no way to make them look alike.

3. If one of the modified expressions is a variable, we check whether the second expression contains the variable. If the variable occurs within the expression, we fail; otherwise, we update our substitution to the composition of the old substitution and a new substitution in which we bind the variable to the second modified expression.

4. The only remaining possibility is that the two modified expressions are both sequences. In this case, we simply iterate across the expressions, comparing as described above.

As an example, consider the computation of the most general unifier for the expressions $p(x,b)$ and $p(a,y)$ with the initial substitution $\{\}$. A trace of the execution of the procedure for this case is shown below. We show the beginning of a comparison with a line labeled Compare together with the expressions being compared and the input substitution. We show the result of each comparison with a line labeled Result. The indentation shows the depth of recursion of the procedure.

$$
\begin{array}{ll}
\text{Compare:} & p(x,b),\, p(a,y),\, \{\,\} \\
& \text{Compare: } p,\, p,\, \{\,\} \\
& \text{Result: } \{\,\} \\
& \text{Compare: } x,\, a,\, \{\,\} \\
& \text{Result: } \{x{\leftarrow}a\} \\
& \text{Compare: } y,\, b,\, \{x{\leftarrow}a\} \\
& \text{Result: } \{x{\leftarrow}a,\, y{\leftarrow}b\} \\
\text{Result:} & \{x{\leftarrow}a,\, y{\leftarrow}b\}
\end{array}
$$

As another example, consider the process of unifying the expression $p(x,x)$ and the expression $p(a,y)$. A trace is shown below. The main interest in this example comes in comparing the last argument in the two expressions, viz. x and y. By the time we reach this point, x is bound to a, so we replace it by a before comparing. y has no binding so we leave it as is. Finally we compare a and y, which results in a binding of y to a.

Compare: $p(x,x), p(a,y), \{\}$
　　　　　　Compare: $p, p, \{\}$
　　　　　　Result: $\{\}$
　　　　　　Compare: $x, a, \{\}$
　　　　　　Result: $\{x \leftarrow a\}$
　　　　　　Compare: $a, y, \{x \leftarrow a\}$
　　　　　　Result: $\{x \leftarrow a, y \leftarrow a\}$
Result:　　　$\{x \leftarrow a, y \leftarrow a\}$

One especially noteworthy part of the unification procedure is the test for whether a variable occurs within an expression before the variable is bound to that expression. This test is called an *occur check* since it is used to check whether or not the variable occurs within the term with which it is being unified. Without this check, the algorithm would find that expressions such as $p(x)$ and $p(f(x))$ are unifiable, even though there is no substitution for x that, when applied to both, makes them look alike.

12.4   RESOLUTION PRINCIPLE

The Relational Resolution Principle is analogous to that of propositional resolution. The main difference is the use of unification to unify literals before applying the rule. Although the rule is simple, there are a couple complexities, so we start with a simple version and then refine it to deal with these complexities.

A simple version of the Resolution Principle for Relational Logic is shown below. Given a clause with a literal φ and a second clause with a literal $\neg\psi$ such that φ and ψ have a most general unifier σ, we can derive a conclusion by applying σ to the clause consisting of the remaining literals from the two original clauses.

$$\frac{\{\varphi_1, \ldots, \varphi, \ldots, \varphi_m\}}{\{\psi_1, \ldots, \neg\psi, \ldots, \psi_n\}}{\{\varphi_1, \ldots, \varphi_m, \psi_1, \ldots, \psi_n\}\sigma}$$
$$\text{where } \sigma = mgu(\varphi, \psi)$$

Consider the example shown below. The first clause contains the positive literal $p(a,y)$ and the second clause contains a negative occurrence of $p(x, f(x))$. The substitution $\{x \leftarrow a, y \leftarrow f(a)\}$ is a most general unifier of these two expressions. Consequently, we can collect the remaining literals $r(y)$ and $q(g(x))$ into a clause and apply the substitution to produce a conclusion.

$$\frac{\{p(a, y), r(y)\}}{\{\neg p(x, f(x)), q(g(x))\}}{\{r(f(a)), q(g(a))\}}$$

Unfortunately, this simple version of the Resolution Principle is not quite good enough. Consider the two clauses shown below. Given the meaning of these two clauses, it should be possible to resolve them to produce the empty clause. However, the two atomic sentences do not unify. The variable x must be bound to a and b at the same time.

$$\{p(a, x)\}$$
$$\{\neg p(x, b)\}$$

Fortunately, this problem can easily be fixed by extending the Resolution Principle slightly as shown below. Before trying to resolve two clauses, we select one of the clauses and rename any variables the clause has in common with the other clause.

$$\{\varphi_1, \dots, \varphi, \dots, \varphi_m\}$$
$$\underline{\{\psi_1, \dots, \neg\psi, \dots, \psi_n\}}$$
$$\{\varphi_1\tau, \dots, \varphi_m\tau, \psi_1, \dots, \psi_n\}\sigma$$

where τ is a variable renaming on $\{\varphi_1, \dots, \varphi, \dots, \varphi_m\}$
where $\sigma = mgu(\varphi\tau, \psi)$

Renaming solves this problem. Unfortunately, we are still not quite done. There is one more technicality that must be addressed to finish the story. As stated, even with the extension mentioned above, the rule is not quite good enough. Given the clauses shown below, we should be able to infer the empty clause $\{\}$; however, this is not possible with the preceding definition. The clauses can be resolved in various ways, but the result is never the empty clause.

$$\{p(x), p(y)\}$$
$$\{\neg p(u), \neg p(v)\}$$

The good news is that we can solve this additional problem with one last modification to our definition of the Resolution Principle. If a subset of the literals in a clause Φ has a most general unifier γ, then the clause Φ' obtained by applying γ to Φ is called a *factor* of Φ. For example, the literals $p(x)$ and $p(f(y))$ have a most general unifier $\{x \leftarrow f(y)\}$, so the clause $\{p(f(y)), r(f(y), y)\}$ is a factor of $\{p(x), p(f(y)), r(x, y)\}$. Obviously, any clause is a trivial factor of itself.

Using the notion of factors, we can give a complete definition for the *Resolution Principle*. Suppose that Φ and Ψ are two clauses. If there is a literal φ in some factor of Φ and a literal $\neg\psi$ in some factor of Ψ, then we say that the two clauses Φ and Ψ *resolve* and that the new clause $((\Phi' - \{\varphi\}) \cup (\Psi' - \{\neg\psi\}))\sigma$ is a *resolvent* of the two clauses.

$$\Phi$$
$$\underline{\Psi}$$
$$((\Phi' - \{\varphi\}) \cup (\Psi' - \{\neg\psi\}))\sigma$$

where τ is a variable renaming on Φ
where Φ' is a factor of $\Phi\tau$ and $\varphi \in \Phi'$
where Ψ' is a factor of Ψ and $\neg\psi \in \Psi'$
where $\sigma = mgu(\varphi, \psi)$

Using this enhanced definition of resolution, we can solve the problem mentioned above. Once again, consider the premises $\{p(x), p(y)\}$ and $\{\neg p(u), \neg p(v)\}$. The first premise has the factor $\{p(x)\}$, and the second has the factor $\{\neg p(u)\}$, and these two factors resolve to the empty clause in a single step.

12.5   RESOLUTION REASONING

Reasoning with the Resolution Principle is analogous to reasoning with the Propositional Resolution Principle. We start with premises; we apply the Resolution Principle to those premises; we apply the rule to the results of those applications; and so forth until we get to our desired conclusion or until we run out of things to do.

As with Propositional Resolution, we define a *resolution derivation* of a conclusion from a set of premises to be a finite sequence of clauses terminating in the conclusion in which each clause is either a premise or the result of applying the Resolution Principle to earlier members of the sequence. And, as with Propositional Resolution, we do not use the word *proof*, because we reserve that word for a slightly different concept, which is discussed in the next section.

As an example, consider a problem in the area kinship relations. Suppose we know that Art is the parent of Bob and Bud; suppose that Bob is the parent of Cal; and suppose that Bud is the parent of Coe. Suppose we also know that grandparents are parents of parents. Starting with these premises, we can use resolution to conclude that Art is the grandparent of Coe. The derivation is shown below. We start with our five premises—four simple clauses for the four facts about the parent relation p and one more complex clause capturing the definition of the grandparent relation g. We start by resolving the clause on line 1 with the clause on line 5 to produce the clause on line 6. We then resolve the clause on line 3 with this result to derive that conclusion that Art is the grandparent of Cal. Interesting but not what we set out to prove; so we continue the process. We next resolve the clause on line 2 with the clause on line 5 to produce the clause on line 8. Then we resolve the clause on line 5 with this result to produce the clause on line 9, which is exactly what we set out to prove.

1.	$\{p(art, bob)\}$	Premise
2.	$\{p(art, bud)\}$	Premise
3.	$\{p(bob, cal)\}$	Premise
4.	$\{p(bud, coe)\}$	Premise
5.	$\{\neg p(x, y), \neg p(y, z), g(x, z)\}$	Premise
6.	$\{\neg p(bob, z), g(art, z)\}$	1, 5
7.	$\{g(art, cal)\}$	3, 6
8.	$\{\neg p(bud, z), g(art, z)\}$	2, 5
9.	$\{g(art, coe)\}$	4, 8

One thing to notice about this derivation is that there are some dead-ends. We first tried resolving the fact about Art and Bob before getting around to trying the fact about Art and Bud. Resolution does not eliminate all search. However, at no time did we ever have to make an arbitrary assumption or an arbitrary choice of a binding for a variable. The absence of such arbitrary choices is why Resolution is so much more focussed than natural deduction systems like Fitch.

Another worthwhile observation about Resolution is that, unlike Fitch, Resolution frequently terminates even when there is no derivation of the desired result. Suppose, for example, we were interested in deriving the clause $\{g(cal,art)\}$ from the premises in this case. This sentence, of course, does *not* follow from the premises. And resolution is sound, so we would never generate this result. The interesting thing is that, in this case, the attempt to derive this result would eventually terminate. With the premises given, there are a few more resolutions we could do, e.g., resolving the clause on line 1 with the second literal in the clause on line 5. However, having done these additional resolutions, we would find ourselves with nothing left to do; and, unlike Fitch, the process would terminate.

Unfortunately, like Propositional Resolution, Resolution is not *generatively complete*, i.e., it is not possible to find resolution derivations for all clauses that are logically entailed by a set of premise clauses. For example, the clause $\{p(a), \neg p(a)\}$ is always true, and so it is logically entailed by any set of premises, including the empty set of premises. Resolution requires some premises to have any effect. Given an empty set of premises, we would not be able to derive any conclusions, including this valid clause.

Although Resolution is not *generatively* complete, problems like this one are solved by negating the goal and demonstrating that the resulting set of sentences is unsatisfiable.

12.6   UNSATISFIABILITY

One common use of resolution is in demonstrating unsatisfiability. In clausal form, a contradiction takes the form of the empty clause, which is equivalent to a disjunction of no literals. Thus, to automate the determination of unsatisfiability, all we need do is to use resolution to derive consequences from the set to be tested, terminating whenever the empty clause is generated.

Let's start with a simple example. See the derivation below. We have four premises. The derivation in this case is particularly easy. We resolve the first clause with the second to get the clause shown on line 5. Next, we resolve the result with the third clause to get the unit clause on line 6. Note that $r(a)$ is the remaining literal from clause 3 after the resolution, and $r(a)$ is also the remaining literal from clause 5 after the resolution. Since these two literals are identical, they appear only once in the result. Finally, we resolve this result with the clause on 4 to produce the empty clause.

1.	$\{p(a,b), q(a,c)\}$	Premise
2.	$\{\neg p(x,y), r(x)\}$	Premise
3.	$\{\neg q(x,y), r(x)\}$	Premise
4.	$\{\neg r(z)\}$	Premise
5.	$\{q(a,c), r(a)\}$	1, 2
6.	$\{r(a)\}$	5, 3
7.	$\{\}$	6, 4

Here is a more complicated derivation, one that illustrates renaming and factoring. Again, we have four premises. Line 5 results from resolution between the clauses on lines 1 and 3. This one is easy. Line 6 results from resolution between the clauses on lines 2 and 4. In this case, renaming is necessary in order for the unification to take place. Line 7 results from renaming and factoring the clause on line 5 and resolving with the clause on line 6. Finally line 8 results from factoring line 5 again and resolving with the clause on line 7. Note that we cannot just factor 5 and factor 6 and resolve the results in one step. Try it and see what happens.

1.	$\{\neg p(x,y), q(x,y,f(x,y))\}$	Premise
2.	$\{r(y,z), \neg q(a,y,z)\}$	Premise
3.	$\{p(x, g(x)), q(x,g(x),z)\}$	Premise
4.	$\{\neg r(x,y), \neg q(x,w,z)\}$	Premise
5.	$\{q(x,g(x),f(x,g(x))), q(x,g(x),z)\}$	1, 3
6.	$\{\neg q(a,x,y), \neg q(x,w,z)\}$	2, 4
7.	$\{\neg q(g(a),w,z)\}$	5, 6 (factoring 5)
8.	$\{\}$	5, 7 (factoring 5)

In demonstrating unsatisfiability, Resolution and Fitch without DC are equally powerful. Given a set of sentences, Resolution can derive the empty clause from the clausal form of the sentences if and only if Fitch can find a proof of a contradiction. The benefit of using Resolution is that the search space is smaller.

12.7   LOGICAL ENTAILMENT

As with Propositional Logic, we can use a test for unsatisfiability to test logical entailment as well. Suppose we wish to show that the set of sentences Δ logically entails the formula φ. We can do this by finding a proof of φ from Δ, i.e., by establishing $\Delta \mid\!- \varphi$. By the refutation theorem, we can establish that $\Delta \mid\!- \varphi$ by showing that $\Delta \cup \{\neg \varphi\}$ is unsatisfiable. Thus, if we show that the set of formulas $\Delta \cup \{\neg \varphi\}$ is unsatisfiable, we have demonstrated that Δ logically entails φ.

To apply this technique of establishing logical entailment by establishing unsatisfiability using resolution, we first negate φ and add it to Δ to yield Δ'. We then convert Δ' to clausal form and apply resolution. If the empty clause is produced, the original Δ' was unsatisfiable, and we have demonstrated that Δ logically entails φ. This process is called a *resolution refutation* ; it is illustrated by examples in the following sections.

As an example of using Resolution to determine logical entailment, let's consider a case we saw earlier. The premises are shown below. We know that everybody loves somebody and everybody loves a lover.

$$\forall x.\exists y.loves(x,y)$$
$$\forall u.\forall v.\forall w.(loves(v,w) \Rightarrow loves(u,v))$$

Our goal is to show that everybody loves everybody.

$$\forall x.\forall y.loves(x,y)$$

In order to solve this problem, we add the negation of our desired conclusion to the premises and convert to clausal form, leading to the clauses shown below. Note the use of a Skolem function in the first clause and the use of Skolem constants in the clause derived from the negated goal.

$$\{loves(x,f(x))\}$$
$$\{\neg loves(v,w), loves(u,v)\}$$
$$\{\neg loves(a,b)\}$$

Starting from these initial clauses, we can use resolution to derive the empty clause and thus prove the result.

1.	$\{loves(x,f(x))\}$	Premise
2.	$\{\neg loves(v,w), loves(u,v)\}$	Premise
3.	$\{\neg loves(a,b)\}$	Premise
4.	$\{loves(u, x)\}$	1, 2
5.	$\{\}$	4, 3

As another example of resolution, once again consider the problem of Harry and Ralph introduced in the preceding chapter. We know that every horse can outrun every dog. Some greyhounds can outrun every rabbit. Greyhounds are dogs. The relationship of being faster is transitive. Harry is a horse. Ralph is a rabbit.

$$\forall x.\forall y.(h(x) \land d(y) \Rightarrow f(x, y))$$
$$\exists y.(g(y) \land \forall z.(r(z) \Rightarrow f(y, z)))$$
$$\forall y.(g(y) \Rightarrow d(y))$$
$$\forall x.\forall y.\forall z.(f(x, y) \land f(y, z) \Rightarrow f(x, z))$$
$$h(harry)$$
$$r(ralph)$$

We desire to prove that Harry is faster than Ralph. In order to do this, we negate the desired conclusion.

$$\neg f(harry, ralph)$$

To do the proof, we take the premises and the negated conclusion and convert to clausal form. The resulting clauses are shown below. Note that the second premise has turned into two clauses.

1.	$\{\neg h(x), \neg d(y), f(x, y)\}$	Premise
2.	$\{g(gary)\}$	Premise
3.	$\{\neg r(z), f(gary, z)\}$	Premise
4.	$\{\neg g(y), d(y)\}$	Premise
5.	$\{\neg f(x, y), \neg f(y, z), f(x, z)\}$	Premise
6.	$\{h(harry)\}$	Premise
7.	$\{r(ralph)\}$	Premise
8.	$\{\neg f(harry, ralph)\}$	Negated Goal

From these clauses, we can derive the empty clause, as shown in the following derivation.

9.	$\{d(gary)\}$	2, 4
10.	$\{\neg d(y), f(harry, y)\}$	6, 1
11.	$\{f(harry, gary)\}$	9, 10
12.	$\{f(gary, ralph)\}$	7, 3
13.	$\{\neg f(gary, z), f(harry, z)\}$	11, 5
14.	$\{f(harry, ralph)\}$	12, 13
15.	$\{\}$	14, 8

Don't be misled by the simplicity of these examples. Resolution can and has been used in proving complex mathematical theorems, in proving the correctness of programs, and in various other applications.

12.8   ANSWER EXTRACTION

In a previous section, we saw how to use resolution in answering true-or-false questions (e.g., *Is Art the grandparent of Coe?*). In this section, we show how resolution can be used to answer fill-in-the-blank questions as well (e.g., *Who is the grandparent of Coe?*).

A fill-in-the-blank question is a sentence with free variables specifying the blanks to be filled in. The goal is to find bindings for the free variables such that the database logically entails the sentence obtained by substituting the bindings into the original question.

For example, to ask about Jon's parent, we would write the question $p(x, jon)$. Using the database from the previous section, we see that *art* is an answer to this question, since the sentence $p(art, jon)$ is logically entailed by the database.

An *answer literal* for a fill-in-the-blank question φ is a sentence $goal(v_1, \dots, v_n)$, where $v_1, \dots, v_n$ are the free variables in φ. To answer φ, we form an implication from φ and its answer literal and convert to clausal form.

For example, the literal $p(x, jon)$ is combined with its answer literal $goal(x)$ to form the rule $(p(x, jon) \Rightarrow goal(x))$, which leads to the clause $\{\neg p(x, jon), goal(x)\}$.

To get answers, we use resolution as described above, except that we change the termination test. Rather than waiting for the empty clause to be produced, the procedure halts as soon as it derives a clause consisting of only answer literals. The following resolution derivation shows how we compute the answer to *Who is Jon's parent?*

1.	$\{f(art, jon)\}$	Premise
2.	$\{f(bob, kim)\}$	Premise
3.	$\{\neg f(x, y), p(x, y)\}$	Premise
4.	$\{\neg p(x, jon), goal(x)\}$	Goal
5.	$\{\neg f(x, jon), goal(x)\}$	3, 4
6.	$\{goal(art)\}$	1, 5

If this procedure produces only one answer literal, the terms it contains constitute the only answer to the question. In some cases, the result of a fill-in-the-blank resolution depends on the refutation by which it is produced. In general, several different refutations can result from the same query, leading to multiple answers.

Suppose, for example, that we knew the identities of both the father and mother of Jon and that we asked *Who is one of Jon's parents?* The following resolution trace shows that we can derive two answers to this question.

1.	$\{f(art, jon)\}$	Premise
2.	$\{m(ann, jon)\}$	Premise
3.	$\{\neg f(x, y), p(x, y)\}$	Premise
4.	$\{\neg m(x, y), p(x, y)\}$	Premise
5.	$\{\neg p(x, jon), goal(x)\}$	Goal
6.	$\{\neg f(x, jon), goal(x)\}$	3, 5
7.	$\{goal(art)\}$	1, 6
8.	$\{\neg m(x, jon), goal(x)\}$	4, 5
9.	$\{goal(ann)\}$	2, 8

Unfortunately, we have no way of knowing whether or not the answer statement from a given refutation exhausts the possibilities. We can continue to search for answers until we find enough of them. However, due to the undecidability of logical entailment, we can never know in general whether we have found all the possible answers.

Another interesting aspect of fill-in-the-blank resolution is that in some cases the procedure can result in a clause containing more than one answer literal. The significance of this is that no one answer is guaranteed to work, but one of the answers must be correct.

The following resolution trace illustrates this fact. The database in this case is a disjunction asserting that either Art or Bob is the father of Jon, but we do not know which man is. The goal is to find a parent of John. After resolving the goal clause with the sentence about fathers and parents, we resolve the result with the database disjunction, obtaining a clause that can be resolved a second time yielding a clause with two answer literals. This answers indicates not two answers but rather uncertainty as to which is the correct answer.

1.	$\{f(art, jon), f(bob, jon)\}$	Premise
2.	$\{\neg f(x, y), p(x, y)\}$	Premise
3.	$\{\neg p(x, jon), goal(x)\}$	Goal
4.	$\{\neg f(x, jon), goal(x)\}$	2, 3
5.	$\{f(art, jon), goal(bob)\}$	1, 4
6.	$\{goal(art), goal(bob)\}$	5, 4

In such situations, we can continue searching in hope of finding a more specific answer. However, given the undecidability of logical entailment, we can never know in general whether we can stop and say that no more specific answer exists.

12.9   STRATEGIES

One of the disadvantages of using the resolution rule in an unconstrained manner is that it leads to many useless inferences. Some inferences are redundant in that their conclusions can be derived in other ways. Some inferences are irrelevant in that they do not lead to derivations of the desired result.

This section presents a number of strategies for eliminating useless work. In reading this material, it is important to bear in mind that we are concerned here not with the order in which inferences are done, but only with the size of a resolution graph and with ways of decreasing that size by eliminating useless deductions.

PURE LITERAL ELIMINATION

A literal occurring in a clause set is *pure* if and only if it has no instance that is complementary to an instance of another literal in the clause set. A clause that contains a pure literal is useless for the purposes of refutation, since the literal can never be resolved away. Consequently, we can safely remove such a clause. Removing clauses with pure literals defines a deletion strategy known as *pure-literal elimination*.

The clause set that follows is unsatisfiable. However, in proving this we can ignore the second and third clauses, since they both contain the pure literal s. The example in this case involves clauses in Propositional Logic, but it applies equally well to Relational Logic.

$$\{\neg p, \neg q, r\}$$
$$\{\neg p, s\}$$
$$\{\neg q, s\}$$
$$\{p\}$$
$$\{q\}$$
$$\{\neg r\}$$

Note that, if a database contains no pure literals, there is no way we can derive any clauses with pure literals using resolution. The upshot is that we do not need to apply the strategy to a database more than once, and in particular we do not have to check each clause as it is generated.

TAUTOLOGY ELIMINATION

A *tautology* is a clause containing a pair of complementary literals. For example, the clause $\{p(f(a)), \neg p(f(a))\}$ is a tautology. The clause $\{p(x), q(y), \neg q(y), r(z)\}$ also is a tautology, even though it contains additional literals.

As it turns out, the presence of tautologies in a set of clauses has no effect on that set's satisfiability. A satisfiable set of clauses remains satisfiable, no matter what tautologies we add. An unsatisfiable set of clauses remains unsatisfiable, even if we remove all tautologies. Therefore, we can remove tautologies from a database, because we need never use them in subsequent inferences. The corresponding deletion strategy is called *tautology elimination*.

Note that the literals in a clause must be exact complements for tautology elimination to apply. We cannot remove non-identical literals, just because they are complements under unification. For example, the clauses $\{\neg p(a), p(x)\}$, $\{p(a)\}$, and $\{\neg p(b)\}$ are unsatisfiable. However, if we were to remove the first clause, the remaining clauses would be satisfiable.

SUBSUMPTION ELIMINATION

In *subsumption elimination*, the deletion criterion depends on a relationship between two clauses in a database. A clause Φ *subsumes* a clause Ψ if and only if there exists a substitution σ such that $\Phi\sigma \subseteq \Psi$. For example, $\{p(x), q(y)\}$ subsumes $\{p(a), q(v), r(w)\}$, since there is a substitution $\{x\leftarrow a, y\leftarrow v\}$ that makes the former clause a subset of the latter.

If one member in a set of clauses subsumes another member, then the set remaining after eliminating the subsumed clause is satisfiable if and only if the original set is satisfiable. Therefore, subsumed clauses can be eliminated. Since the resolution process itself can produce tautologies and subsuming clauses, we need to check for tautologies and subsumptions as we perform resolutions.

UNIT RESOLUTION

A *unit resolvent* is one in which at least one of the parent clauses is a *unit clause*, i.e., one containing a single literal. A *unit derivation* is one in which all derived clauses are unit resolvents. A *unit refutation* is a unit derivation of the empty clause.

As an example of a unit refutation, consider the following proof. In the first two inferences, unit clauses from the initial set are resolved with binary clauses to produce two new unit clauses. These are resolved with the first clause to produce two additional unit clauses. The elements in these two sets of results are then resolved with each other to produce the contradiction.

1.	$\{p, q\}$	Premise
2.	$\{\neg p, r\}$	Premise
3.	$\{\neg q, r\}$	Premise
4.	$\{\neg r\}$	Premise
5.	$\{\neg p\}$	2, 4
6.	$\{\neg q\}$	3, 4
7.	$\{q\}$	1, 5
8.	$\{p\}$	1, 6
9.	$\{r\}$	3, 7
10.	$\{\}$	6, 7

Note that the proof contains only a subset of the possible uses of the resolution rule. For example, clauses 1 and 2 can be resolved to derive the conclusion $\{q, r\}$. However, this conclusion and its descendants are never generated, since neither of its parents is a unit clause.

Inference procedures based on unit resolution are easy to implement and are usually quite efficient. It is worth noting that, whenever a clause is resolved with a unit clause, the conclusion has fewer literals than the parent does. This helps to focus the search toward producing the empty clause and thereby improves efficiency.

Unfortunately, inference procedures based on unit resolution generally are not complete. For example, the clauses $\{p, q\}$, $\{\neg p, q\}$, $\{p, \neg q\}$, and $\{\neg p, \neg q\}$ are inconsistent. Using general resolution, it is easy to derive the empty clause. However, unit resolution fails in this case, since none of the initial clauses contains just one literal.

On the other hand, if we restrict our attention to Horn clauses (i.e., clauses with at most one positive literal), the situation is much better. In fact, it can be shown that there is a unit refutation of a set of Horn clauses if and only if it is unsatisfiable.

INPUT RESOLUTION

An *input resolvent* is one in which at least one of the two parent clauses is a member of the initial (i.e., input) database. An *input deduction* is one in which all derived clauses are input resolvents. An *input refutation* is an input deduction of the empty clause.

It can be shown that unit resolution and input resolution are equivalent in inferential power in that there is a unit refutation from a set of sentences whenever there is an input refutation and vice versa.

One consequence of this fact is that input resolution is complete for Horn clauses but incomplete in general. The unsatisfiable set of clauses $\{p, q\}$, $\{\neg p, q\}$, $\{p, \neg q\}$, and $\{\neg p, \neg q\}$ provides an example of a deduction on which input resolution fails. An input refutation must (in particular) have one of the parents of $\{\}$ be a member of the initial database. However, to produce the empty clause in this case, we must resolve either two single literal clauses or two clauses having single-literal factors. None of the members of the base set meet either of these criteria, so there cannot be an input refutation for this set.

LINEAR RESOLUTION

Linear resolution (also called *ancestry-filtered resolution*) is a slight generalization of input resolution. A *linear resolvent* is one in which at least one of the parents is either in the initial database or is an ancestor of the other parent. A *linear deduction* is one in which each derived clause is a linear resolvent. A *linear refutation* is a linear deduction of the empty clause.

Linear resolution takes its name from the linear shape of the proofs it generates. A linear deduction starts with a clause in the initial database (called the *top clause*) and produces a linear chain of resolution. Each resolvent after the first one is obtained from the last resolvent (called the *near parent*) and some other clause (called the *far parent*). In linear resolution, the far parent must either be in the initial database or be an ancestor of the near parent.

Much of the redundancy in unconstrained resolution derives from the resolution of intermediate conclusions with other intermediate conclusions. The advantage of linear resolution is that it avoids many useless inferences by focusing deduction at each point on the ancestors of each clause and on the elements of the initial database.

Linear resolution is known to be refutation complete. Furthermore, it is not necessary to try every clause in the initial database as top clause. It can be shown that, if a set of clauses Δ is satisfiable and $\Delta \cup \{\Phi\}$ is unsatisfiable, then there is a linear refutation with Φ as top clause. So, if we know that a particular set of clauses is consistent, we need not attempt refutations with the elements of that set as top clauses.

A *merge* is a resolvent that inherits a literal from each parent such that this literal is collapsed to a singleton by the most general unifier. The completeness of linear resolution is preserved even if the ancestors that are used are limited to merges. Note that, in this example, the first resolvent (i.e., clause $\{q\}$) is a merge.

SET OF SUPPORT RESOLUTION

If we examine resolution traces, we notice that many conclusions come from resolutions between clauses contained in a portion of the database that we know to be satisfiable. For example, in many cases, the set of premises is satisfiable, yet many of the conclusions are obtained by resolving premises with other premises rather than the negated conclusion As it turns out, we can eliminate these resolutions without affecting the refutation completeness of resolution.

A subset Γ of a set Δ is called a *set of support* for Δ if and only if $\Delta - \Gamma$ is satisfiable. Given a set of clauses Δ with set of support Γ, a *set of support resolvent* is one in which at least one parent is selected from Γ or is a descendant of Γ. A *set of support deduction* is one in which each derived clause is a set of support resolvent. A *set of support refutation* is a set of support deduction of the empty clause.

The following derivation is a set of support refutation, with the singleton set $\{\neg r\}$ as the set of support. The clause $\{\neg r\}$ resolves with $\{\neg p, r\}$ and $\{\neg q, r\}$ to produce $\{\neg p\}$ and $\{\neg q\}$. These then resolve with clause 1 to produce $\{q\}$ and $\{p\}$, which resolve to produce the empty clause.

1.	$\{p, q\}$	Premise
2.	$\{\neg p, r\}$	Premise
3.	$\{\neg q, r\}$	Premise
4.	$\{\neg r\}$	Set of Support
5.	$\{\neg p\}$	2, 4
6.	$\{\neg q\}$	3, 4
7.	$\{q\}$	1, 5
8.	$\{\}$	7, 6

Obviously, this strategy would be of little use if there were no easy way of selecting the set of support. Fortunately, there are several ways this can be done at negligible expense. For example, in situations where we are trying to prove conclusions from a consistent database, the natural choice is to use the clauses derived from the negated goal as the set of support. This set satisfies the definition as long as the database itself is truly satisfiable. With this choice of set of support, each resolution must have a connection to the overall goal, so the procedure can be viewed as working backward from the goal. This is especially useful for databases in which the number of conclusions possible by working forward is larger. Furthermore, the goal-oriented character of such refutations often makes them more understandable than refutations using other strategies.

RECAP

The *Resolution Principle* is a rule of inference for Relational Logic analogous to the Propositional Resolution Principle for Propositional Logic. As with Propositional Resolution, Resolution works only on expressions in *clausal form*. Unification is the process of determining whether two expressions can be *unified*, i.e., made identical by appropriate substitutions for their variables. A *substitution* is a finite mapping of variables to terms. The variables being replaced together constitute the *domain* of the substitution, and the terms replacing them constitute the *range*. A substitution is *pure* if and only if all replacement terms in the range are free of the variables in the domain of the substitution. Otherwise, the substitution is *impure*. The result of applying a substitution σ to an expression φ is the expression $\varphi\sigma$ obtained from the original expression by replacing every occurrence of every variable in the domain of the substitution by the term with which it is associated. The *composition* of two substitutions is a single substitution that has the same effect as applying those substitutions in sequence. A substitution σ is a *unifier* for an expression φ and an expression ψ if and only if $\varphi\sigma=\psi\sigma$, i.e., the result of applying σ to φ is the same as the result of applying σ to ψ. If two expressions have a unifier, they are said to be *unifiable*. Otherwise, they are *nonunifiable*. A *most general unifier*, or *mgu*, σ of two expressions has the property that it is as general as or more general than any other unifier. Although it is possible for two expressions to have more than one most general unifier, all of these most general unifiers are structurally the same, i.e., they are unique up to variable renaming. The Resolution Principle is analogous to that of Propositional Resolution. The main difference is the use of unification to unify literals before applying the rule. A *resolution derivation* of a conclusion from a set of premises is a finite sequence of clauses terminating in the conclusion in which each clause is either a premise or the result of applying the Resolution Principle to earlier members of the sequence. Resolution and Fitch without DC are equally powerful. Given a set of sentences, Resolution can derive the empty clause from the clausal form of the sentences if and only if Fitch can find a proof of a contradiction. The benefit of using Resolution is that the search space is smaller.

12.10 EXERCISES

12.1. Consider a language with two object constants a and b and one function constant f. Give the clausal form for each of the following sentences in this language.

(a) $\exists y.\forall x.p(x,y)$
(b) $\forall x.\exists y.p(x,y)$
(c) $\exists x.\exists y.(p(x,y) \wedge q(x,y))$
(d) $\forall x.\forall y.(p(x,y) \Rightarrow q(x))$
(e) $\forall x.(\exists y.p(x,y) \Rightarrow q(x))$

12.2. For each of the following pairs of sentences, say whether the sentences are unifiable and give a most general unifier for those that are unifiable.

(a) $p(x,x)$ and $p(a,y)$
(b) $p(x,x)$ and $p(f(y),z)$
(c) $p(x,x)$ and $p(f(y),y)$
(d) $p(f(x,y),g(z,z))$ and $p(f(f(w,z),v),w)$

12.3. Give all resolvents, if any, for each of the following pairs of clauses.

(a) $\{p(x,f(x)), q(x)\}$ and $\{\neg p(a,y), r(y)\}$
(b) $\{p(x,b), q(x)\}$ and $\{\neg p(a,x), r(x)\}$
(c) $\{p(x), p(a), q(x)\}$ and $\{\neg p(y), r(y)\}$
(d) $\{p(x), p(a), q(x)\}$ and $\{\neg p(y), r(y)\}$
(e) $\{p(a), q(y)\}$ and $\{\neg p(x), \neg q(b)\}$
(f) $\{p(x), q(x,x)\}$ and $\{\neg q(a, f(a))\}$

12.4. Given the clauses $\{p(a), q(a)\}, \{\neg p(x), r(x)\}, \{\neg q(a)\}$, use Resolution to derive the clause $\{r(a)\}$.

12.5. Given the premises $\forall x.(p(x) \Rightarrow q(x))$ and $\forall x.(q(x) \Rightarrow r(x))$, use Resolution to prove the conclusion $\forall x.(p(x) \Rightarrow r(x))$.

12.6. Given $\forall x.(p(x) \Rightarrow q(x))$, use Resolution to prove $\forall x.p(x) \Rightarrow \forall x.q(x)$.

12.7. Use Resolution to prove $\forall x.(((p(x) \Rightarrow q(x)) \Rightarrow p(x)) \Rightarrow p(x))$.

12.8. Given $\forall x.\forall y.\forall z.(p(x,y) \wedge p(y,z) \Rightarrow p(x,z))$, $\forall x.p(x,a)$, and $\forall y.p(a,y)$, use Resolution to prove $\forall x.\forall y.p(x,y)$.

12.9. Use resolution to show that the clauses $\{\neg p(x,y), q(x,y,f(x,y))\}$, $\{\neg r(y,z), q(a,y,z)\}$, $\{r(y,z), \neg q(a,y,z)\}$, $\{p(x,g(x)), q(x,g(x),z)\}$, and $\{\neg r(x,y), \neg q(x,w,z)\}$ are unsatisfiable. This one is a little tricky. Be careful about factoring.

12.10. Given $p(a)$ and $\forall x.(p(x) \Rightarrow q(x) \vee r(x))$, use Answer Extraction to find a τ such that $(q(\tau) \vee r(\tau))$ is true.

Bibliography

[1] David Barker-Plummer, Jon Barwise, and John Etchemendy. *CSLI Publications*, Stanford, CA, 2nd ed.

[2] Armin Biere, Marijn J. H. Heule, Hans van Maaren, and Toby Walsh, Eds. *Handbook of Satisfiability*, volume 185 of Frontiers in Artificial Intelligence and Applications. IOS Press, Lansdale, PA, February 2009.

[3] Hans K. Buning and T. Letterman. *Propositional Logic: Deduction and Algorithms*. Cambridge University Press, New York, NY, 1999.

[4] Chin-Liang Chang and Richard C. Lee. *Symbolic Logic and Mechanical Theorem Proving (Computer Science Classics)*. Academic Press, Salt Lake City, UT, May 1973. DOI: 10.1016/c2009-0-22103-9.

[5] Herbert B. Enderton. *A Mathematical Introduction to Logic*. Academic Press, Salt Lake City, UT, 2nd ed., 2001.

[6] Timothy Hinrichs and Michael Genesereth. Herbrand Logic. Technical Report LG-2006-02, Stanford University, Stanford, CA, 2006. http://logic.stanford.edu/reports/LG-2006-02.pdf.

[7] John Alan Robinson and Andrei Voronkov, Eds. *Handbook of Automated Reasoning*, (in 2 volumes). MIT Press, Cambridge, MA, 2001.

Authors' Biographies

MICHAEL GENESERETH

Michael Genesereth is an associate professor in the Computer Science Department at Stanford University. He received his Sc.B. in Physics from M.I.T. and his Ph.D. in Applied Mathematics from Harvard University. He is best known for his research on Computational Logic and its applications. He has been teaching Logic to Stanford students and others for more than 20 years. He is the current director of the Logic Group at Stanford and founder and research director of CodeX (The Stanford Center for Legal Informatics).

ERIC J. KAO

Eric J. Kao is a member of technical staff at VMware, Inc. He received his B.Math in Pure Mathematics from the University of Waterloo and his Ph.D. in Computer Science from Stanford University. His work centers on Computational Logic and its applications in cloud computing, cybersecurity, and data management. He currently leads OpenStack Congress, a leading open-source collaboration for declarative cloud management. He is passionate about education and making technical subjects more accessible. In his free time, he tutors and mentors high school students in East Palo Alto.